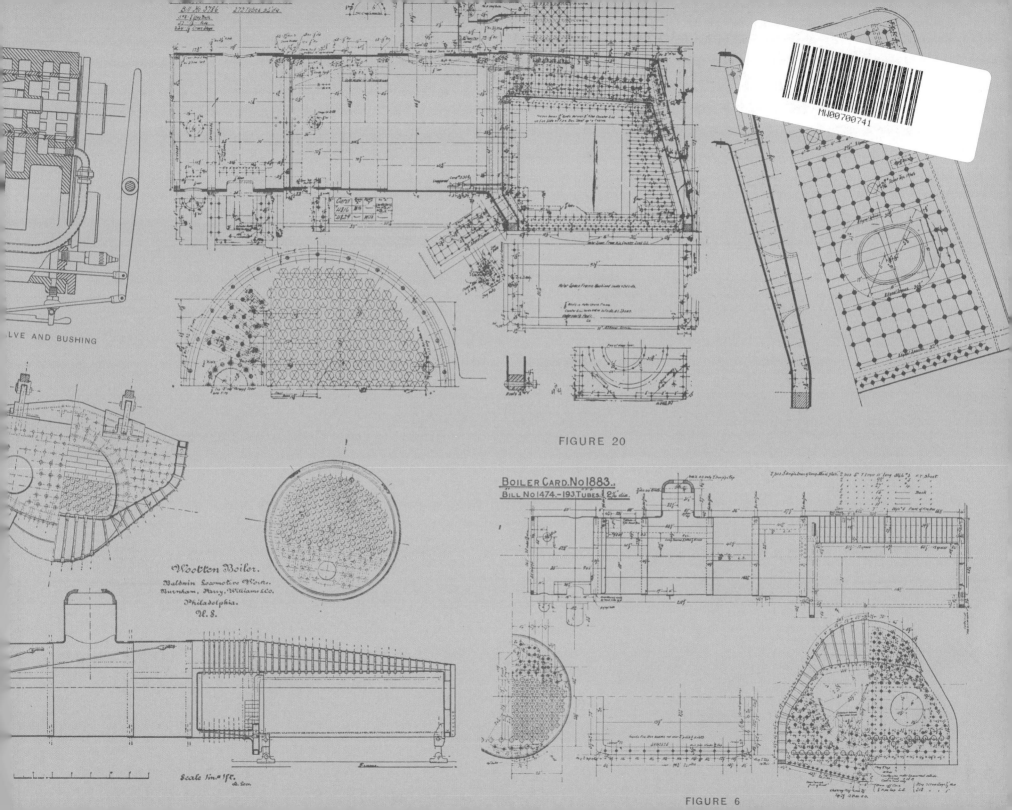

LVE AND BUSHING

FIGURE 20

Wootten Boiler.
Baldwin Locomotive Works.
Burnham, Parry, Williams & Co.
Philadelphia.
U.S.

Scale 1 in.= 1 ft.

BOILER CARD.No 1883.
BILL No 1474.- 193 TUBES. 2¼ dia.

FIGURE 6

RECORD OF RECENT CONSTRUCTION

NOS. 21 TO 30 INCLUSIVE

BALDWIN LOCOMOTIVE WORKS

BURNHAM, WILLIAMS & CO.

PRINCIPAL OFFICE, 500 NORTH BROAD STREET

PHILADELPHIA, U.S.A.

4880 Lower Valley Road, Atglen, Pennsylvania 19310

Baldwin Locomotive Works

RECENT years have shown such rapid advances in locomotive practice, both by the adoption of larger locomotives and by improvements in details of construction, that it has been difficult to prepare a general illustrated catalogue which would not become obsolete almost as soon as issued. It has, therefore, been deemed desirable to publish particulars of current construction, without attempting to formulate a complete scheme of types and sizes of locomotives. The following pages contain information pertaining to a great variety of locomotives, of different gauges, and for different kinds of service representing current requirements. They are presented without special arrangement and mainly in the order of construction. These pages are also used to present other matters of interest in connection with locomotive work.

The total number of locomotives built up to and including the month of February, 1902, is 20,000. The annual capacity of the works is 1500 locomotives. The present organization, based on this capacity, is as follows :

Number of men employed	11,000
Hours of labor per man per day	10
Principal departments run continuously, hours per day	23
Horse-power employed	7,000
Number of buildings comprised in the works	33
Acreage comprised in the works	16
Number of dynamos for furnishing power	9
Horse-power of dynamos for furnishing power	1,800
Number of dynamos for furnishing light (arc)	8
Number of dynamos for furnishing light (incandescent)	5
Horse-power of electric motors employed for power transmission, aggregate	3,500
Number of electric lamps in service (incandescent)	3,500
Number of electric lamps in service (arc)	400
Consumption of coal, in net tons, per week, approximately	2,150
Consumption of iron, in net tons, per week, approximately	3,500
Consumption of other materials, in net tons, per week, approximately	1,000

Schiffer Books are available at special discounts for bulk purchases for sales promotions or premiums. Special editions, including personalized covers, corporate imprints, and excerpts can be created in large quantities for special needs. For more information contact the publisher:

Published by Schiffer Publishing Ltd.
4880 Lower Valley Road
Atglen, PA 19310
Phone: (610) 593-1777; Fax: (610) 593-2002
E-mail: Info@schifferbooks.com

For the largest selection of fine reference books on this and related subjects, please visit our web site at **www.schifferbooks.com**
We are always looking for people to write books on new and related subjects. If you have an idea for a book please contact us at the above address.

This book may be purchased from the publisher.
Include $5.00 for shipping.
Please try your bookstore first.
You may write for a free catalog.

In Europe, Schiffer books are distributed by
Bushwood Books
6 Marksbury Ave.
Kew Gardens
Surrey TW9 4JF England
Phone: 44 (0) 20 8392 8585; Fax: 44 (0) 20 8392 9876
E-mail: info@bushwoodbooks.co.uk
Website: www.bushwoodbooks.co.uk

VICTORIAN RAILWAYS, AUSTRALIA

Test Train, Nyora to Melbourne, 781 Tons behind Tender, consisting of 54 vehicles, including Car and Van.   Ruling gradient 1 in 75.

BALDWIN LOCOMOTIVE WORKS

Compound Consolidation Locomotive

Class 10 $\frac{20}{38}$-E-40

for the

Victorian Railways, Australia

Gauge 5' 3"

GENERAL DIMENSIONS

CYLINDERS

Diameter (High Pressure) . .	13"
" (Low Pressure) . .	22"
Stroke	26"
Valve . . .	Balanced Piston

BOILER

Diameter	56"
Thickness of Sheets . . .	⅝"
Working Pressure . .	200 lbs.
Fuel	Soft Coal

FIREBOX

Material	Copper
Length	85"
Width	47"
Depth, Front . . .	55½"
" Back . . .	53¼"
Thickness of Sheets, Sides .	½"
" " " Back . .	½"
" " " Crown .	½"
" " " Tube	⅞" and ½"

TUBES

Number	198
Diameter	2"
Length	13' 1½"

HEATING SURFACE

Firebox	111 sq. ft.
Tubes . . .	1347 sq. ft.
Total	1458 sq. ft.
Grate Area . . .	27.74 sq. ft.

DRIVING WHEELS

Diameter Outside . . .	54"
" of Center . .	48"
Journals . . .	7½" x 9"

ENGINE TRUCK WHEELS

Diameter	30"
Journals	5" x 8"

WHEEL BASE

Driving	14' 0"
Total Engine . . .	21' 11"
Total Engine and Tender . .	48' 6"

WEIGHT

On Drivers . . .	106,800 lbs.
On Truck . . .	14,200 lbs.
Total Engine . . .	121,000 lbs
Total Engine and Tender .	193,000 lbs

TENDER

Diameter of Wheels . .	33
Journals	4¼" x 8"
Tank Capacity . .	3,600 gal.

SERVICE

Freight.

To traverse easily curves of eight chains (528 feet) radius from main line to sidings. To haul 2,700 tons on a level.

BALDWIN LOCOMOTIVE WORKS

Compound Ten-Wheel Locomotive

Class 10 $\frac{25}{46}$-D-6

for the

Gauge 4' 8½"

Chicago, Rock Island & Pacific Railway Company

GENERAL DIMENSIONS

CYLINDERS

Diameter (High Pressure) .	15½"
" (Low Pressure) . .	26"
Stroke	28"
Valve . . .	Balanced Piston

BOILER

Diameter	66"
Thickness of Sheets .	11/16" and ¾"
Working Pressure . .	200 lbs.
Fuel	Soft Coal

FIREBOX

Material	Steel
Length	118"
Width	40⅛"
Depth, Front . . .	79½"
" Back	67"
Thickness of Sheets, Sides .	5/16"
" " " Back . .	⅜"
" " " Crown .	⅜"
" " " Tube . .	⅝"

TUBES

Number	329
Diameter	2"
Length	15' 0"

HEATING SURFACE

Firebox . . .	180.5 sq. ft.
Tubes	2569.6 sq. ft.
Total	2750.1 sq. ft.
Grate Area . . .	32.8 sq. ft.

DRIVING WHEELS

Diameter Outside . . .	78½"
" of Center . .	72"
Journals . . .	9" x 12"

ENGINE TRUCK WHEELS

Diameter	36"
Journals . . .	6½" x 11"

WHEEL BASE

Driving	14' 6"
Total Engine . . .	26' 9"
Total Engine and Tender .	53' 6⅝"

WEIGHT

On Drivers . . .	134,560 lbs.
On Truck	44,715 lbs.
Total Engine . .	179,275 lbs.
Total Engine and Tender .	290,000 lbs.

TENDER

Diameter of Wheels . .	36"
Journals . . .	5" x 9"
Tank Capacity . .	5,500 gals.
	10 Tons Coal

SERVICE

Passenger.

BALDWIN LOCOMOTIVE WORKS

Compound Consolidation Locomotive

Class 10 $\frac{25}{46}$-E-107

for the

Gauge 4' 8½"

Grand Trunk Railway Company

GENERAL DIMENSIONS

CYLINDERS

Diameter (High Pressure) .	15½"
" (Low Pressure) . .	26"
Stroke	28"
Valve . . .	Balanced Piston

BOILER

Diameter	66"
Thickness of Sheets . .	11⁄16"
Working Pressure . .	200 lbs.
Fuel	Soft Coal

FIREBOX

Material	Steel
Length	114¾6"
Width	41⅜"
Depth, Front . . .	65½"
" Back . . .	62½"
Thickness of Sheets, Sides .	5⁄16"
" " " Back .	5⁄16"
" " " Crown .	⅜"
" " " Tube .	½"

TUBES

Number	260
Diameter	2"
Length	14' 0"

HEATING SURFACE

Firebox . . .	161.1 sq. ft.
Tubes . . .	1894.4 sq. ft.
Total . . .	2055.5 sq. ft.
Grate Area . .	32.7 sq. ft.

DRIVING WHEELS

Diameter Outside . . .	56"
" of Center . .	50"
Journals . . .	8" x 12"

ENGINE TRUCK WHEELS

Diameter	33"
Journals	5" x 9½"

WHEEL BASE

Driving . . .	15' 3"
Total Engine . . .	23' 6"
Total Engine and Tender .	53' 9"

WEIGHT

On Drivers . . .	143,400 lbs.
On Truck . . .	17,900 lbs.
Total Engine . .	161,300 lbs.
Total Engine and Tender .	251,000 lbs.

TENDER

Diameter of Wheels . .	33"
Journals	5" x 9"
Tank Capacity . .	4,500 gals.

SERVICE

Freight.

BALDWIN LOCOMOTIVE WORKS

Ten-Wheel Passenger Locomotive

Class 10-32-D-641

for the

Gauge 4' 9"

Paris & Orleans Railway Company

GENERAL DIMENSIONS

CYLINDERS

Diameter	19"
Stroke	26"
Valve	Balanced Piston

BOILER

Diameter	60"
Thickness of Sheets	11⁄16"
Working Pressure	213 lbs.
Fuel	Coal

FIREBOX

Material	Copper
Length	92"
Width	41½"
Depth, Front	67½"
" Back	55"
Thickness of Sheets, Sides	⅝"
" " " Back	⅝"
" " " Crown	⅝"
" " " Tube	⅝ and 1³⁄16"

TUBES

Number	228
Diameter	2"
Length	14' 3"

HEATING SURFACE

Firebox	132 sq. ft.
Tubes	1682.5 sq. ft.
Total	1814.5 sq. ft.
Grate Area	27 sq. ft.

DRIVING WHEELS

Diameter Outside	68.9"
" of Center	1.610 mm.
Journals	8" x 10"

ENGINE TRUCK WHEELS

Diameter	33½"
Journals	5½" x 10"

WHEEL BASE

Driving	13' 4"
Rigid	13' 4"
Total Engine	24' 11"
Total Engine and Tender	47' 4½"

WEIGHT

On Drivers	101,140 lbs.
On Truck	33,585 lbs.
Total Engine	134,725 lbs.
Total Engine and Tender	222,000 lbs.

TENDER

Diameter of Wheels	1.270 mm.
Journals	130 mm. x 240 mm.
Tank Capacity	4,490 gal.
Coal Capacity	9,000 lbs.

SERVICE

Passenger

BALDWIN LOCOMOTIVE WORKS

Consolidation Locomotive

for the

West Virginia Central & Pittsburg Railway Company

Class 10-38-E-173

Gauge 4′ 8½″

GENERAL DIMENSIONS

CYLINDER

Diameter	22″
Stroke	28″
Valve	Balanced

BOILER

Diameter	76″
Thickness of Sheets	¾″
Working Pressure	190″
Fuel	Soft Coal

FIREBOX

Material	Steel
Length	125¹¹⁄₁₆″
Width	42⅝″
Depth, Front	71½″
" Back	67½″
Thickness of Sheets, Sides	⁵⁄₁₆″
" " " Back	⁵⁄₁₆″
" " " Crown	⅜″
" " " Tube	½″

TUBES

Number	300
Diameter	2¼″
Length	13′ 10″

HEATING SURFACE

Firebox	200 sq. ft.
Tubes	2,430 sq. ft.
Total	2,630 sq. ft.
Grate Area	37 sq. ft.

DRIVING WHEELS

Diameter Outside	51″
" of Center	44″
Journals	9″ x 12″

ENGINE TRUCK WHEELS

Diameter	30″
Journals	6″ x 10″

WHEEL BASE

Driving	14′ 3″
Total Engine	22′ 10″
Total Engine and Tender	52′ 9½″

WEIGHT

On Drivers	154,200 lbs.
On Truck	12,800 lbs.
Total Engine	167,000 lbs.
Total Engine and Tender	286,000 lbs.

TENDER

Diameter of Wheels	33″
Journals	5″ x 9″
Tank Capacity	6,000 gals.

SERVICE

Freight.

BALDWIN LOCOMOTIVE WORKS

Compound Ten-Wheel Locomotive

Class 10 $\frac{22}{42}$-D-163

for the

Iowa Central Railway Company

Gauge 4' 8½"

GENERAL DIMENSIONS

CYLINDERS

Diameter (High Pressure) . . .	14"
" (Low Pressure) . .	24"
Stroke	26"
Valve	Balanced Piston

BOILER

Diameter	60"
Thickness of Sheets . .	11/16"
Working Pressure . .	200 lbs.
Fuel	Soft Coal

FIREBOX

Material	Steel
Length	103⅛"
Width	42"
Depth, Front . . .	69"
" Back . . .	66½"
Thickness of Sheets, Sides .	⅜"
" " " Back .	⅜"
" " " Crown .	½"
" " " Tube .	½"

TUBES

Number	264
Diameter	2"
Length	15' 0"

HEATING SURFACE

Firebox	163 sq. ft.
Tubes	2063 sq. ft.
Total	2226 sq. ft.
Grate Area . . .	30 sq. ft.

DRIVING WHEELS

Diameter Outside . . .	62"
" of Center . .	56"
Journals . . .	8½" x 12"

ENGINE TRUCK WHEELS

Diameter	30"
Journals . . .	5½" x 10"

WHEEL BASE

Driving	13' 6"
Total Engine . . .	25' 3"
Total Engine and Tender .	52' 0"

WEIGHT

On Drivers . . .	111,900 lbs.
On Truck . . .	41,250 lbs.
Total Engine . .	153,150 lbs.
Total Engine and Tender	243,000 lbs.

TENDER

Diameter of Wheels . .	33"
Journals . . .	5" x 9"
Tank Capacity . .	4,500 gals.

SERVICE

Freight.

BALDWIN LOCOMOTIVE WORKS

Consolidation Locomotive

for the

Finland State Railways

Class 10-26-E-281

Gauge 5′ 0″

GENERAL DIMENSIONS

CYLINDERS

Diameter	16″
Stroke	20″
Valve	Balanced

BOILER

Diameter	48″
Thickness of Sheets . . .	½″
Working Pressure . . .	180″
Fuel	Coal

FIREBOX

Material	Copper
Length	60¾″
Width	36¼″
Depth, Front . . .	53″
" Back . . .	46¼″
Thickness of Sheets, Sides . .	½″
" " " Back . .	½″
" " " Crown .	½″
" " " Tube 7/8″ and ½″	

TUBES

Number	130
Diameter	2″
Length	11′ 10″

HEATING SURFACE

Firebox . . .	70.62 sq. ft.
Tubes . . .	796.21 sq. ft.
Total . . .	866.83 sq. ft.
Grate Area . . .	14.8 sq. ft.

DRIVING WHEELS

Diameter Outside . . .	44.1″
" of Center . .	38.975″
Journal . . .	6″ x 8″

ENGINE TRUCK WHEELS

Diameter	31¼″
Journals . . .	4½″ x 7½″

WHEEL BASE

Driving	12′ 6″
Total Engine . . .	19′ 0″
Total Engine and Tender .	37′ 0½″

WEIGHT

On Drivers . . .	71,905 lbs.
On Truck . . .	10,600 lbs.
Total Engine . . .	82,505 lbs.
Total Engine and Tender	122,000 lbs.

TENDER

Diameter of Wheels . .	37¼″
Journals	4″ x 8″
Tank Capacity . .	2,000 gals.

SERVICE

Freight

BALDWIN LOCOMOTIVE WORKS

Six-Coupled Double-Ender Locomotive

Class 12-24¼-D-1

for the

Gauge 5' 0"

Finland State Railways

GENERAL DIMENSIONS

CYLINDERS

Diameter	15"
Stroke	24"
Valve	Balanced

BOILER

Diameter	50"
Thickness of Sheets	½"
Working Pressure	180 lbs.
Fuel	Coal

FIREBOX

Material	Copper
Length	56¾"
Width	37"
Depth, Front	57"
" Back	43¾"
Thickness of Sheets, Sides	½"
" " " Back	½"
" " " Crown	½"
" " " Tube	⅞" and ½"

TUBES

Number	157
Diameter	2"
Length	10' 9⅜"

HEATING SURFACE

Firebox	70.1 sq. ft.
Tubes	874.6 sq. ft.
Total	944.7 sq. ft.
Grate Area	14.6 sq. ft.

DRIVING WHEELS

Diameter Outside	49"
" of Center	44"
Journals	6½" x 8"

ENGINE TRUCK WHEELS

Diameter, Front and Back	31¼"
Journals	4½" x 7½"

WHEEL BASE

Driving	10' 3"
Total Engine	28' 6"

WEIGHT

On Drivers	62,900 lbs.
On Truck, Front	13,600 lbs.
" " Back	29,500 lbs.
Total Engine	106,000 lbs.
Tank Capacity	1,500 gals.

SERVICE

Freight.

BALDWIN LOCOMOTIVE WORKS

Six-Coupled Switching Locomotive

for the

Riverside Iron Works

Class 6-30-D-277

Gauge 4' 8½"

GENERAL DIMENSIONS

CYLINDERS

Diameter . . . . . 18"
Stroke . . . . . 24"
Valve . . . . . Balanced

BOILER

Diameter . . . . . 54"
Thickness of Sheets . . . ⁹⁄₁₆"
Working Pressure . . . 150 lbs.
Fuel . . . . . Soft Coal

FIREBOX

Material . . . . . Steel
Length . . . . . 69⅜"
Width . . . . . 42"
Depth, Front . . . 54¾"
"   Back . . . 53¼"
Thickness of Sheets, Sides . . ⅜"
"   "   "   Back . . ⅜"
"   "   "   Crown . ½"
"   "   "   Tube . ½"

TUBES

Number 193
Diameter . . . . . 2"
Length . . . . . 12' 4"

HEATING SURFACE

Firebox . . . . 94.39 sq. ft.
Tubes . . . . 1237.88 sq. ft.
Total . . . . 1332.27 sq. ft.
Grate Area . . . . 20.23 sq. ft.

DRIVING WHEELS

Diameter Outside . . . 44"
"   of Center . . 38"
Journals . . . . 8" x 8½"

WHEEL BASE

Driving . . . . . 11' 0"
Total . . . . . 11' 0"

WEIGHT

On Drivers . . . 114,570 lbs.
Total . . . . 114,570 lbs.
Tank Capacity . . 1,500 gals.

SERVICE

Switching.

BALDWIN LOCOMOTIVE WORKS

Four-Coupled Double-Ender Locomotive

for the

Class 8-26¼-C-2   Gauge 4' 8½"

Hastings Lumber Company

GENERAL DIMENSIONS

CYLINDERS

Diameter	16"
Stroke	22"
Valve	Balanced

BOILER

Diameter	52"
Thickness of Sheets	½"
Working Pressure	160 lbs.
Fuel	Soft Coal

FIREBOX

Material	Steel
Length	60⁷⁄₁₆"
Width	34⅜"
Depth, Front	58"
" Back	56½"
Thickness of Sheets, Sides	⁵⁄₁₆"
" " " Back	⁵⁄₁₆"
" " " Crown	⅜"
" " " Tube	½"

TUBES

Number	175
Diameter	2"
Length	10' 10"

HEATING SURFACE

Firebox	77.7 sq. ft.
Tubes	984.3 sq. ft.
Total	1062.0 sq. ft.
Grate Area	14.4 sq. ft.

DRIVING WHEELS

Diameter Outside	44"
" of Center	38"
Journals	7½" x 8½"

ENGINE TRUCK WHEELS

(Front and Back)

Diameter	26"
Journals	5" x 8"

WHEEL BASE

Driving	7' 0"
Total Engine	22' 8"

WEIGHT

On Drivers	94,500 lbs.
On Truck, Front	13,200 lbs.
" " Back	14,000 lbs.
Total	121,700 lbs.
Tank Capacity	
	1,000 gals. on sides of boiler

SERVICE

Logging

BALDWIN LOCOMOTIVE WORKS

Mogul Locomotive

Class 8-26-D-112

for the

Gauge 5′ 6″

Oudh & Rohilkund Railway (Indian State Railways)

GENERAL DIMENSIONS

CYLINDERS

Diameter	16″
Stroke	24″
Valve	Balanced

BOILER

Diameter	54″
Thickness of Sheets . .	9/16″
Working Pressure . .	160 lbs.
Fuel	Soft Coal

FIREBOX

Material	Copper
Length	55¾″
Width	43″
Depth	68″
Thickness of Sheets, Sides .	½″
" " " Back . .	½″
" " " Crown .	½″
" " " Tube	⅞″ and ½″

TUBES

Number	186
Diameter	2″
Length	10′ 8⅝″

HEATING SURFACE

Firebox	102.2 sq. ft.
Tubes	1025.3 sq. ft.
Total	1127.5 sq. ft.
Grate Area . . .	16.64 sq. ft.

DRIVING WHEELS

Diameter Outside . . .	54¾″
" of Center . .	49½″
Journals	7″ x 8″

ENGINE TRUCK WHEELS

Diameter	30″
Journals	5″ x 8″

WHEEL BASE

Driving	13′ 11″
Total Engine	21′ 3″
Total Engine and Tender .	40′ 10½″

WEIGHT

On Drivers . . .	77,000 lbs.
On Truck	18,300 lbs.
Total Engine . . .	95,300 lbs.
Total Engine and Tender .	137,000 lbs.

TENDER

Diameter of Wheels . . .	43″
Journals	4½″ x 9″
Tank Capacity . . .	2,160 gals.

SERVICE

Freight.

BALDWIN LOCOMOTIVE WORKS

Four-Coupled Locomotive

Class 6-11-C-8

for the

Gauge 3' 0''

Gray Lumber Company

GENERAL DIMENSIONS

CYLINDERS

Diameter	9''
Stroke	14''
Valve	Plain

BOILER

Diameter	32''
Thickness of Sheets	3/8''
Working Pressure	160 lbs.
Fuel	Wood

FIREBOX

Material	Steel
Length	43 5/16''
Width	24 1/8''
Depth	31 1/2''
Thickness of Sheets, Sides	5/16''
" " " Back	5/16''
" " " Crown	5/16''
" " " Tube	3/8''

TUBES

Number	79
Diameter	1 1/2''
Length	6' 0''

HEATING SURFACE

Firebox	35.0 sq. ft.
Tubes	183.5 sq. ft.
Total	218.5 sq. ft.
Grate Area	7.0 sq. ft.

DRIVING WHEELS

Diameter of Outside	33''
" of Center	28''
Journals	4 1/2'' x 6''

ENGINE TRUCK WHEELS

Diameter	22''
Journals	3'' x 5''

WHEEL BASE

Driving	5' 6''
Total Engine	11' 2''
Total Engine and Tender	25' 7''

WEIGHT

On Drivers	23,580 lbs.
On Truck	3,700 lbs.
Total Engine	27,280 lbs.
Total Engine and Tender	37,000 lbs.

TENDER

Diameter of Wheels	22''
Journals	2 3/4'' x 5''
Tank Capacity	400 gals.

SERVICE

Switching

BALDWIN LOCOMOTIVE WORKS

Six-Coupled Locomotive

Class 6-18-D-18

for the

Gauge 3′ 3⅜″

Mijnbouw Maatschappij Herrero, Spain

GENERAL DIMENSIONS

CYLINDERS

Diameter	12″
Stroke	18″
Valve	Balanced

BOILER

Diameter	38″
Thickness of Sheets	7⁄16″
Working Pressure	160 lbs.
Fuel	Coal

FIREBOX

Material	Copper
Length	45¼″
Width	26½″
Depth	40″
Thickness of Sheets, Sides	½″
" " " Back	½″
" " " Crown	½″
" " " Tube	¾″ and ½″

TUBES

Number	72
Diameter	2″
Length	9′ 0″

HEATING SURFACE

Firebox	43.43 sq. ft.
Tubes	334.96 sq. ft.
Total	378.59 sq. ft.
Grate Area	8.3 sq. ft.

DRIVING WHEELS

Diameter Outside	38″
" of Center	32″
Journals	5½ x 7″

WHEEL BASE

Driving	8′ 6″
Total Engine	8′ 6″

WEIGHT

On Drivers	48,180 lbs.
Total Engine	48,180 lbs.
Tank Capacity	500 gals.

SERVICE

Switching

BALDWIN LOCOMOTIVE WORKS

Eight-Coupled Double-Ender Locomotive

Class 12-20¼-E-3

for the

Gauge 2' 5½"

Jokkis Forssa Railway Company (Finland)

GENERAL DIMENSIONS

CYLINDERS

Diameter	13"
Stroke	16"
Valve	Plain

BOILER

Diameter	36"
Thickness of Sheets . . .	⅜"
Working Pressure . . .	160 lbs.
Fuel	Coal

FIREBOX

Material	Copper
Length	37"
Width	39"
Depth	43½"
Thickness of Sheets, Sides . .	½"
" " " Back .	½"
" " " Crown . .	½"
" " " Tube	¾" and ½"

TUBES

Number	93
Diameter	1¾"
Length	11' 9"

HEATING SURFACE

Firebox	44 sq. ft.
Tubes	495.9 sq. ft.
Total	539.9 sq. ft.
Grate Area	10 sq. ft.

DRIVING WHEELS

Diameter Outside	30"
" of Center . . .	25"
Journals	5" x 6"

ENGINE TRUCK WHEELS
(Front and Back)

Diameter	23"
Journals	4" x 6"

WHEEL BASE

Driving	8' 6"
Total	21' 2"

WEIGHT

On Drivers	43,600 lbs
On Truck, Front . .	8,500 lbs.
" " Back . .	10,200 lbs.
Total Engine . . .	62,300 lbs.
Tank Capacity . . .	850 gals.

SERVICE

Freight.

Minimum curves to be 100 meters. Grades 1.6 per cent. (84½ feet per mile). Length of road, 100 kilometers. Maximum speed, 30 kilometers per hour.

BALDWIN LOCOMOTIVE WORKS

Compound Double Locomotive

Class 12 $\frac{17}{32}$-DD-1

for the

McCloud River Railroad Company

Gauge 4′ 8½″

GENERAL DIMENSIONS

CYLINDERS

(Two Sets)

Diameter (High Pressure) .	11½″
" (Low Pressure) . .	19″
Stroke	20″
Valve . . .	Balanced Piston

BOILERS

Diameter	46″
Thickness of Sheets . . .	½″
Working Pressure . .	200 lbs.
Fuel	Wood

FIREBOXES

Material	Steel
Length	53¹¹⁄₁₆″
Width	34⅜″
Depth	59″
Thickness of Sheets, Sides .	⁵⁄₁₆″
" " " Back .	⁵⁄₁₆″
" " " Crown .	⅜″
" " " Tube .	½″

TUBES

Number in each boiler . .	136
Diameter	2″
Length	12′ 9″

HEATING SURFACE

(Aggregate)

Fireboxes	148 sq. ft.
Tubes	1804 sq. ft.
Total	1952 sq. ft.
Grate Area	26 sq. ft.

DRIVING WHEELS

Diameter Outside . . .	40″
" of Center . . .	34″
Journals	6½″ x 8″

WHEEL BASE

Rigid	9′ 9″
Total	38′ 4″

WEIGHT

On Drivers . . .	161,400 lbs.
Total	161,400 lbs.
Tank Capacity (aggregate)	3 400 gals.

SERVICE

Freight.

Guaranteed to haul 125 tons of 2,000 lbs. up a grade of 7%.

BALDWIN LOCOMOTIVE WORKS

THE LOCOMOTIVE shown on page (32) was designed with a view of meeting the same conditions of traffic as are now performed by the well-known Mallet Articulated Locomotives, the Meyer Double Bogie Locomotives, and the Fairlie Locomotives. The object in view was to obtain simplicity of construction, cheapen the first cost or selling price, and at the same time obtain maximum efficiency.

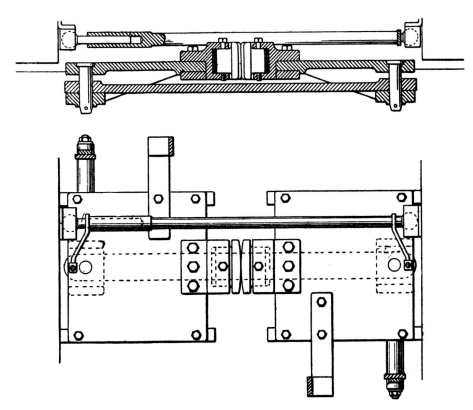

Description

It will be observed that the locomotive is composed of two separate and exactly similar engines connected end to end at the firebox or back bumper. The medium is merely the ordinary draw-bar connection commonly used between engines and tenders, except that it is made of great strength compared to the work required of it.

The holes have clearance space next the draw pins to permit the strains to pass through the buffers when the engine is used as a pusher, and small bearing strips are provided, as shown by accompanying illustration, immediately above the bar to prevent unusual tipping of the engines at their point of connection.

Throttle Rigging

It is natural to wonder how this device is arranged in order to enable the engineer to control both engines from one position.

34

Referring to the illustration on page herewith, it will be observed that the throttle levers are made of the crank pattern, opening as they are pulled to the right or toward the engineer. Two levers are provided for each engine. One is fast to the throttle rod of its engine, and the other is connected across to the other engine by various rods and other mechanism, so adjusted that the position of the lever or the length of its connections is not materially altered, no matter what position the engines may assume on the track in relation to each other. With this arrangement the two levers may be used separately by the engineer, or they may be locked together and used as a single lever, opening or closing the throttles of both engines simultaneously, thus giving the engineer perfect control of both engines while looking in the direction in which the locomotive is moving.

Reverse Lever

The reversing mechanism, as shown on the following page, has also been arranged in duplicate, that is, one lever on each engine always coupled to the lever of the other engine. Unlike the throttle they cannot be

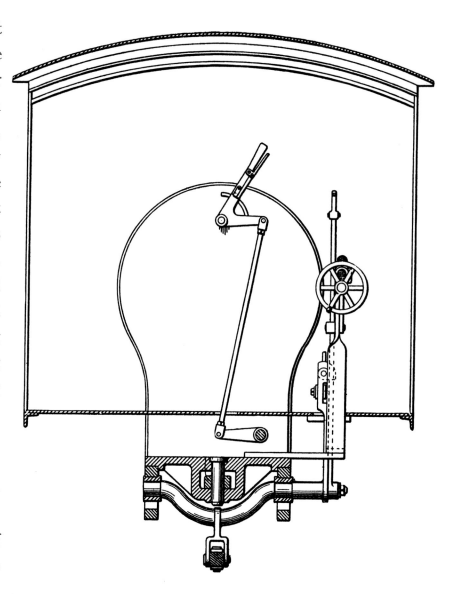

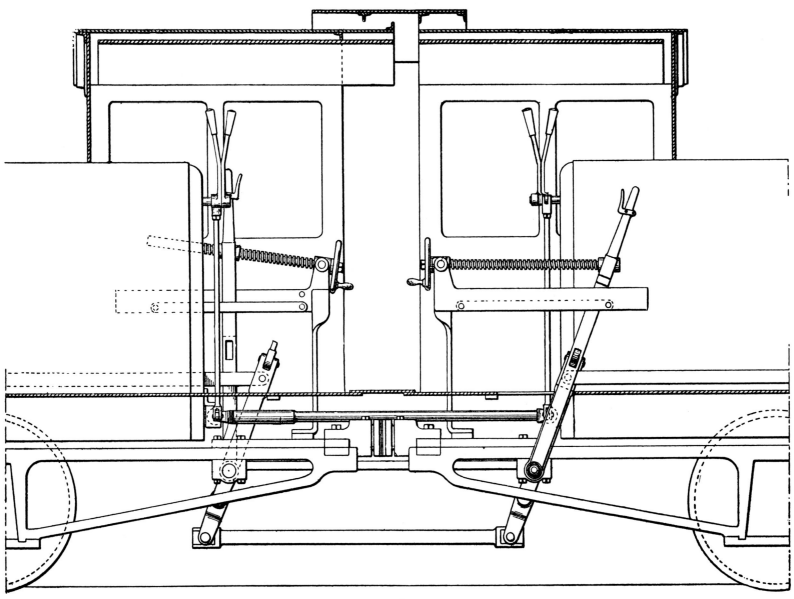

handled separately, provision being made to hold up the latch of the lever on the rear or unattended engine, so that it can be operated by the lever of the other engine as shown. The connecting device between the two engines for the reverse levers is so arranged as to be centrally parallel with, and below the draw-bar, and of exactly the same length. By this means the action is positive and the valve gear is not affected by the curving of the locomotive.

Cabs

Protection for the engine crew is provided for in the regular manner, by placing a cab on the back end of each engine and providing a covering over their roofs where they join to prevent entrance of rain and covering the connection of their floors by a sheet of iron. In case of necessity, the sides can be connected by a closed connection similar to car vestibules.

Fuel

The fuel is carried on the left side of each engine, and if the fuel is wood, it can be arranged as shown, but if coal is used, the water capacity can be increased and sufficient fuel carried in suitable bunkers placed ahead of back end of boiler on left side of each engine.

Advantages

It is sometimes desirable to use such a locomotive when the traffic is heavy, the grades and curves severe, and the line difficult to operate, but after improvements in the track and roadbed, such a locomotive

BALDWIN LOCOMOTIVE WORKS

may become undesirable. With any of the types heretofore used, a change of type is impossible, but with this locomotive by merely removing the draw-bar pins and connections between throttle levers and reverse levers, two separate locomotives are at once obtained, either of which can be operated independently, as a shunter or if provided with a tender, at small cost can be converted into a road locomotive.

In the Duplex locomotive there are no flexible steam pipe joints to be taken care of, the path of the steam is as direct as in the ordinary locomotives and not subject to the same conditions as in other types, thus avoiding much condensation and loss of fuel.

Outside steam pipes and outside throttles are avoided.

Owing to the flexibility of the connection there is less resistance to curvature than in any other of the three types mentioned. The Meyer and Fairlie locomotives with their double trucks, and the Mallet locomotive with one rigid and one hinged framing, do not give equal flexibility on curves.

The complication of the Mallet, the Meyer, or the Fairlie, as noticed in the drawings and illustrations of these engines, conduces to greater cost of repairs, greater loss of time in service, and consequently increased capital invested in rolling stock.

The first cost of Duplex locomotives is also much less than that of either of the other types mentioned, the parts are more thoroughly interchangeable and thus require fewer spare parts to insure constant service from the locomotive.

If at any time either section of the locomotive becomes disabled from any cause, requiring repairs extending over a long period, the other section or half can be separated from the injured section and used as a separate locomotive, doing half the work, thus maintaining traffic which would cease altogether in case the locomotive had been a Fairlie, a Meyer, or a Mallet.

BALDWIN LOCOMOTIVE WORKS

An advantage which the Duplex locomotive possesses in common with the Fairlie, over the Meyer and Mallet, is the large grate and heating service obtainable, which should be an important item to purchasers when considering the various types of double bogie or articulated locomotives.

The practicability of this type of locomotive has been fully demonstrated, together with its great advantages for moving maximum loads over light rails, poor roadbed, and through sharp curvatures.

BALDWIN LOCOMOTIVE WORKS

Compound Atlantic Type Locomotive

Class 10 $\frac{24}{14}$ ¼-C-7

for the

Gauge 4' 8½"

Baltimore & Ohio Railroad Company

GENERAL DIMENSIONS

CYLINDERS

Diameter (High Pressure) . .	15"
" (Low Pressure) . .	25"
Stroke	28"
Valve . . .	Balanced Piston

BOILER

Diameter	62"
Thickness of sheets .	11/16" and 5/8"
Working Pressure . .	200 lbs.
Fuel	Soft Coal

FIREBOX

Material	Steel
Length	101 15/16"
Width	60 1/8"
Depth, Front . . .	64"
" Back . . .	62"
Thickness of Sheets, Sides .	5/16"
" " " Back .	5/16"
" " " Crown .	3/8"
" " " Tube .	1/2"

TUBES

Number . . .	300
Diameter . . .	2"
Length . . .	16' 1'

HEATING SURFACE

Firebox . . .	150 sq. ft.
Tubes . . .	2513 sq. ft.
Total . . .	2663 sq. ft.
Grate Area . .	42.5 sq. ft.

DRIVING WHEELS

Diameter Outside . .	78"
" of Center . .	72"
Journals . .	8½" x 12"

ENGINE TRUCK WHEELS

Diameter . . .	33"
Journals . .	5½" x 10"

TRAILING WHEELS

Diameter . . .	48"
Journals . .	8½" x 12"

WHEEL BASE

Driving . . .	6' 9"
Rigid . . .	13' 6"
Total Engine . .	25' 7"
Total Engine and Tender .	52' 6½"

WEIGHT

On Drivers . .	83,400 lbs.
On Truck . .	37,940 lbs.
On Trailing Wheels .	28,260 lbs.
Total Engine . .	149,600 lbs.
Total Engine and Tender	249,000 lbs.

TENDER

Diameter of Wheels . .	36"
Journals . . .	5" x 9"
Tank Capacity . .	5,000 gals.

SERVICE

Fast Passenger.

BALDWIN LOCOMOTIVE WORKS

Compound Consolidation Locomotive

for the

Baltimore & Ohio Railroad Company

Class 10 $\frac{25}{46}$-E-80

Gauge 4' 8½"

GENERAL DIMENSIONS

CYLINDERS

Diameter (High Pressure) .	15½"
" (Low Pressure) .	26"
Stroke	30"
Valve . . .	Balanced Piston

BOILER

Diameter	64"
Thickness of Sheets .	11⁄16" and ¾"
Working Pressure .	200 lbs.
Fuel	Soft Coal

FIREBOX

Material	Steel
Length	118"
Width	41⅛"
Depth, Front . . .	70⅛"
" Back . . .	67⅛"
Thickness of Sheets, Sides .	5⁄16"
" " " Back .	⅜"
" " " Crown .	⅜"
" " " Tube .	½"

TUBES

Number	246
Diameter	2¼"
Length . . .	14' 10½"

HEATING SURFACE

Firebox	179.5 sq. ft.
Tubes . . .	2142.9 sq. ft.
Fire Brick Tubes . .	25.1 sq. ft.
Total . . .	2347.5 sq. ft.
Grate Area . .	33.7 sq. ft.

DRIVING WHEELS

Diameter Outside . .	54"
" of Center . .	48"
Journals . . .	9" x 10"

ENGINE TRUCK WHEELS

Diameter	30"
Journals . . .	5" x 10"

WHEEL BASE

Driving	15' 4"
Total Engine . . .	23' 8"
Total Engine and Tender .	52' 7"

WEIGHT

On Drivers . . .	163,300 lbs.
On Truck . . .	19,000 lbs.
Total Engine . . .	182,300 lbs.
Total Engine and Tender .	282,000 lbs.

TENDER

Diameter of Wheels . .	33"
Journals . . .	4¼" x 8"
Tank Capacity . .	5,000 gals.

SERVICE

Freight.

BALDWIN LOCOMOTIVE WORKS

Compound Consolidation Locomotive

for the

Baltimore & Ohio Railroad Company

Class 10 $\frac{25}{16}$-E-140   Gauge 4' 8½"

GENERAL DIMENSIONS

CYLINDERS

Diameter (High Pressure) . 15½"
" (Low Pressure) . . 26"
Stroke . . . . . 30"
Valve . . . Balanced Piston

BOILER

Diameter . . . . . 70"
Thickness of Sheets . ¾" and 25⁄32"
Working Pressure . . 200 lbs.
Fuel . . . . Soft Coal

FIREBOX

Material . . . . Steel
Length . . . . . 114"
Width . . . . . 96"
Depth, Front . . . . 63"
" Back . . . . 54"
Thickness of Sheets, Sides . 5⁄16"
" " " Back . ⅜"
" " " Crown . ⅜"
" " " Tube . ½"

TUBES

Number 247
Diameter . . . . 2¼"
Length . . . . 14' 10½"

HEATING SURFACE

Firebox . . . 185.7 sq. ft.
Tubes . . . 2148.3 sq. ft.
Total . . . 2334 sq. ft.
Grate Area . . . 76 sq. ft.

DRIVING WHEELS

Diameter Outside . . . 54"
" of Center . . 48"
Journals . . . 9" x 10"

ENGINE TRUCK WHEELS

Diameter . . . . . 30"
Journals . . . . 5" x 10"

WHEEL BASE

Driving . . . . 15' 4"
Total Engine . . . 23' 8"
Total Engine and Tender . 52' 7"

WEIGHT

On Drivers . . . 166,954 lbs.
On Truck . . . 19,550 lbs.
Total Engine . . 186,504 lbs.
Total Engine and Tender 286,000 lbs.

TENDER

Diameter of Wheels . . . 33"
Journals . . . 4¼" x 8"
Tank Capacity . . . 5,000 gals.

SERVICE

Freight.

45

K.BAY.STS.B.

2398

BALDWIN LOCOMOTIVE WORKS

Compound Atlantic Type Locomotive

Class 10 $\frac{20}{38}$ ¼-C-55

for the

Gauge 4' 8½"

Bavarian State Railways

GENERAL DIMENSIONS

CYLINDERS

Diameter (High Pressure) . .	13"
" (Low Pressure) . .	22"
Stroke	26"
Valve . . .	Balanced Piston

BOILER

Diameter	60"
Thickness of Sheets . ⅝" and 11⁄16"
Working Pressure . . .	200 lbs.
Fuel	Soft Coal

FIREBOX

Material	Steel
Length	103⅛"
Width	42"
Depth, Front	67½"
" Back	65"
Thickness of Sheets, Sides . .	⅜"
" " " Back .	⅜"
" " " Crown .	½"
" " " Tubes .	½"

TUBES

Number	264
Diameter	2"
Length	15' 0"

HEATING SURFACE

Firebox . . .	148.3 sq. ft.
Tubes . . .	2061.8 sq. ft.
Total . . .	2210.1 sq. ft.
Grate Area . . .	30.4 sq. ft.

DRIVING WHEELS

Diameter Outside . . .	72"
" of Center . .	66"
Journals	8" x 12"

ENGINE TRUCK WHEELS

Diameter	33"
Journals . . .	5½" x 10"

TRAILING WHEELS

Diameter	48"
Journals . . .	8" x 12"

WHEEL BASE

Driving	6' 9"
Rigid	13' 9"
Total Engine . . .	25' 5"
Total Engine and Tender .	51' 11"

WEIGHT

On Drivers . . .	68,440 lbs.
On Truck . .	35,020 lbs.
On Trailing Wheels .	29,250 lbs.
Total Engine . .	132,710 lbs.
Total Engine and Tender .	242,000 lbs.

TENDER

Diameter of Wheels . .	36"
Journals	5" x 9"
Tank Capacity . .	5,550 gals.

SERVICE

Fast Passenger.

BALDWIN LOCOMOTIVE WORKS

Compound Ten-Wheel Locomotive

Class 10 $\frac{25}{46}$-D-30

for

Gauge 4' 8½''

The Union Pacific Railroad Company

GENERAL DIMENSIONS

CYLINDERS

Diameter (High Pressure)	15½''
" (Low Pressure)	26''
Stroke	28''
Valve	Balanced Piston

BOILER

Diameter	66''
Thickness of Sheets	11/16 and ¾''
Working Pressure	200 lbs.
Fuel	Soft Coal

FIREBOX

Material	Steel
Length	118 3/16''
Width	39 1/8''
Depth, Front	79½''
" Back	67''
Thickness of Sheets, Sides	5/16''
" " " Back	5/16''
" " " Crown	3/8''
" " " Tube	½''

TUBES

Number	350
Diameter	2''
Length	15' 6''

HEATING SURFACE

Firebox	186 sq. ft.
Tubes	2825 sq. ft.
Total	3011 sq. ft.
Grate Area	32 sq. ft.

DRIVING WHEELS

Diameter Outside	69''
" of Center	62''
Journals	9'' x 12''

ENGINE TRUCK WHEELS

Diameter	30''
Journals	6½'' x 11''

WHEEL BASE

Driving	14' 6''
Total Engine	26' 9''
Total Engine and Tender	54' 0''

WEIGHT

On Drivers	142,440 lbs.
On Truck	41,800 lbs.
Total Engine	184,240 lbs.
Total Engine and Tender	292,000 lbs.

TENDER

Diameter of Wheels	33''
Journals	5'' x 9''
Tank Capacity	6,000 gals.

SERVICE

Passenger.

49

BALDWIN LOCOMOTIVE WORKS

Compound Mogul Locomotive

Class 8 $\frac{25}{46}$-D-16

for the

Gauge 4′ 8½″

Union Pacific Railroad Company (Oregon Short Line)

GENERAL DIMENSIONS

CYLINDERS

Diameter (High Pressure) .	15½″
" (Low Pressure) . .	26″
Stroke	28″
Valve . . .	Balanced Piston

BOILER

Diameter	72″
Thickness of Sheets . .	¾″
Working Pressure . .	200 lbs.
Fuel	Soft Coal

FIREBOX

Material	Steel
Length	120³⁄₁₆″
Width	40⅜″
Depth, Front . . .	71¼″
" Back . . .	69″
Thickness of Sheets, Sides .	⁵⁄₁₆″
" " " Back .	⁵⁄₁₆″
" " " Crown	⅜″
" ' " Tube .	½″

TUBES

Number . . .	298
Diameter . . .	2″
Length . . .	12′ 8″

HEATING SURFACE

Firebox . . .	192 sq. ft.
Tubes . . .	1962 sq. ft.
Total . . .	2154 sq. ft.
Grate Area . . .	33.6 sq. ft.

DRIVING WHEELS

Diameter Outside . . .	57″
" of Center . .	50″
Journals . . .	9″ x 12″

ENGINE TRUCK WHEELS

Diameter	30″
Journals . . .	5½″ x 10″

WHEEL BASE

Driving	15′ 0″
Total Engine . . .	23′ 5″
Total Engine and Tender .	52′ 4½″

WEIGHT

On Drivers . . .	148,160 lbs.
On Truck . . .	20,700 lbs.
Total Engine . . .	168,860 lbs.
Total Engine and Tender .	269,000 lbs.

TENDER

Diameter of Wheels . . .	33″
Journals	5″ x 9″
Tank Capacity . .	5,000 gals.

SERVICE

Freight.

BALDWIN LOCOMOTIVE WORKS

Compound Consolidation Locomotive

Class 10 $\frac{25}{46}$-E-297

for the

Gauge 4' 8½"

Union Pacific Railroad Company

GENERAL DIMENSIONS

CYLINDERS

Diameter (High Pressure)	15½"
" (Low Pressure)	26"
Stroke	30"
Valve	Balanced Piston

BOILER

Diameter	72"
Thickness of Sheets	11/16" and ¾"
Working Pressure	200 lbs.
Fuel	Soft Coal

FIREBOX

Material	Steel
Length	120³⁄16"
Width	40⅝"
Depth, Front	74¼"
" Back	71¼"
Thickness of Sheets, Sides	5/16"
" " " Back	5/16"
" " " Crown	⅜"
" " " Tube	½"

TUBES

Number	321
Diameter	2"
Length	13' 6"

HEATING SURFACE

Firebox	194.9 sq. ft.
Fire Brick Tubes	25.6 sq. ft.
Tubes	2255 sq. ft.
Total	2475.5 sq. ft.
Grate Area	33.8 sq. ft.

DRIVING WHEELS

Diameter Outside	57"
" of Center	50"
Journals	9" x 10"

ENGINE TRUCK WHEELS

Diameter	30"
Journals	6" x 10"

WHEEL BASE

Driving	15' 3"
Total Engine	23' 11"
Total Engine and Tender	53' 5¾"

WEIGHT

On Drivers	161,020 lbs.
On Truck	24,300 lbs.
Total Engine	185,320 lbs.
Total Engine and Tender	305,000 lbs.

TENDER

Diameter of Wheels	33"
Journals	5" x 9"
Tank Capacity	6,000 gals.

SERVICE

Freight.
Radius of Curves 10°.

Code Word, RECHTUIT

BALDWIN LOCOMOTIVE WORKS

Ten-Wheel Locomotive

Class 10-34-D-433

for the

Gauge 4′ 8½″

New York Central & Hudson River Railroad Company

GENERAL DIMENSIONS

CYLINDERS

Diameter	20″
Stroke	28″
Valve	Balanced

BOILERS

Diameter	67 5/16″
Thickness of Sheets, 21/32″, 11/16″ and ¾″	
Working Pressure	200 lbs.
Fuel	Soft Coal

FIREBOX

Material	Steel
Length	108⅛″
Width	40⅜″
Depth, Front	80⅝″
" Back	67⅝″
Thickness of Sheets, Sides	5/16″
" " " Back	⅜″
" " " Crown	⅜″
" " " Tube	½″

TUBES

Number	366
Diameter	2″
Length	14′ 4″

HEATING SURFACE

Firebox	184.3 sq. ft.
Tubes	2730.7 sq. ft.
Total	2915 sq. ft.
Grate Area	30.5 sq. ft.

DRIVING WHEELS

Diameter Outside	75″
" of Center	68″
Journals	9″ x 12″

ENGINE TRUCK WHEELS

Diameter	36″
Journals	6¼″ x 10″

WHEEL BASE

Driving	14′ 11″
Total Engine	26′ 0″
Total Engine and Tender	53′ 0½″

WEIGHT

On Drivers	128,805 lbs.
On Truck	40,800 lbs.
Total Engine	169,605 lbs.
Total Engine and Tender	269,000 lbs.

TENDER

Diameter of Wheels	36″
Journals	5″ x 9″
Tank Capacity	5,000 gals.

SERVICE

Passenger.

BALDWIN LOCOMOTIVE WORKS

Compound Decapod Locomotive

for the

Class 12 $\frac{28}{50}$-F-1

Minneapolis, St. Paul & Sault Ste. Marie Railway Company

Gauge 4' 8½"

GENERAL DIMENSIONS

CYLINDERS

Diameter (High Pressure) .	17"
" (Low Pressure) .	28"
Stroke	32"
Valve . . .	Balanced Piston

BOILER

Diameter	68"
Thickness of Sheets .	11/16" and ¾"
Working Pressure . .	215 lbs.
Fuel . . .	Soft Coal.

FIREBOX

Material	Steel
Length	131 15/16"
Width	41 1/8"
Depth, Front . .	77 3/8"
" Back . . .	76"
Thickness of Sheets, Sides .	5/16"
" " " Back .	5/16"
" " " Crown	3/8"
" " " Tube	½"

TUBES

Number	344
Diameter . . .	2"
Length . . .	15' 7"

HEATING SURFACE

Firebox . . .	223.9 sq. ft.
Tubes . . .	2791.8 sq. ft.
Total . . .	3015.7 sq. ft.
Grate Area . .	37.5 sq. ft.

DRIVING WHEELS

Diameter Outside . .	55"
" of Center . .	48"
Journals, Main . .	9½" x 12"
" Others . .	8½" x 12"

ENGINE TRUCK WHEELS

Diameter . . .	30"
Journals . . .	6" x 10"

WHEEL BASE

Driving . . .	19' 4"
Total Engine . .	28' 0"
Total Engine and Tender .	57' 4"

WEIGHT

On Drivers . .	184,360 lbs.
On Truck . . .	22,850 lbs.
Total Engine . .	207,210 lbs.
Total Engine and Tender .	327,000 lbs.

TENDER

Diameter of Wheels . .	33"
Journals . . .	5½" x 10"
Tank Capacity . .	7,000 gals.
" " .	9 tons Coal

SERVICE

Guaranteed to haul a train weighing 2,000 tons (of 2,000 pounds) exclusive of engine and tender up a grade of 42 feet per mile at a speed of six miles per hour; curve resistance not taken into consideration.

TEN-WHEEL LOCOMOTIVE WITH VANDERBILT BOILER
BUILT AT THE SHOPS OF THE NEW YORK CENTRAL AND HUDSON RIVER R. R. CO.

BALDWIN LOCOMOTIVE WORKS

Locomotive Boilers

A Paper read by Cornelius Vanderbilt, M. E., before the meeting of the Junior Members of the

American Society of Mechanical Engineers

January 8, 1901

IT is not within the scope of this paper to enter into the subject of locomotive boilers to any greater extent than to attempt to show the line along which the boiler has developed, and to give special attention to one of the most recent designs. It is safe to say that the locomotive boiler has changed less since its inception than any other equally important invention, for in looking back at its history we find after the first few locomotives, a general type of boiler adopted which has been followed, with but few variations, up to the present time It is difficult to say exactly why this has been the case, even making due allowance for the peculiar conditions which limit such designs, and the few developments and improvements which have come have followed such very narrow lines that it would almost seem as though there must be some insuperable difficulties in the way of any progress in this direction; that is to say, the majority of the attempts to improve have taken the form of different arrangements for staying a flat-sided firebox or a partially curved crown, as but few have attempted to alter certain well established shapes for the fireboxes themselves. If we classify all boilers under two general heads, namely: External and Internal Firebox Boilers, we find that practically all locomotive boilers—even from the very earliest ones—come under the same head. It is true that a few attempts have been made to design external firebox locomotive boilers, but they have met with but little success, owing principally to the increased cost of maintenance, due primarily to the constant renewals of firebrick, which must of necessity be employed in any such design.

The first locomotive proper which came into use was Richard Trevithick's engine, in 1804, the boiler of which had a cylindrical cast iron shell with a wrought iron internal cylindrical return flue. It is obvious that no large amount of heating surface can be obtained in any such design, and therefore, but few improvements in boilers were made until the Multitubular System was introduced. This was adopted by Stephenson, in the boiler of the "Rocket" locomotive, designed and built by him in 1829. The front portion of the boiler through which the tubes ran, was cylindrical and three feet four inches in diameter; the tubes were six feet long and three inches outside diameter, the firebox was rectangular, projecting behind the rear end of the cylindrical portion of the boiler and was three feet wide and two feet long; it was surrounded on three sides and on the top by a water-leg three inches wide.

The Multitubular System is so largely the basis of every modern locomotive boiler outside of the firebox section, that it is interesting to note its origin. In 1826 a Mr. Neville took out the English patent for a vertical tubular boiler, which stated that the system was equally applicable to horizontal boilers. It would therefore seem as though we owe this portion of the modern design to

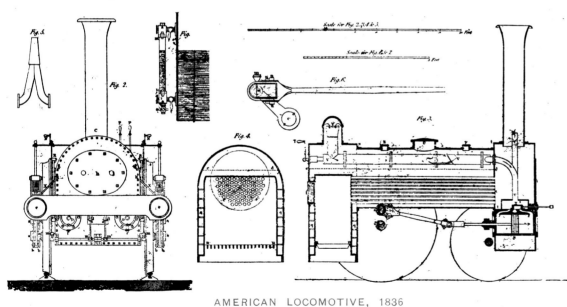

AMERICAN LOCOMOTIVE, 1836

BALDWIN LOCOMOTIVE WORKS

the vertical land boiler, but this however, is not entirely undisputed, as the
idea is often credited to a Mr. Booth of
the Liverpool and Manchester Railway.
Stephenson improved his original
"Rocket" boiler in 1833, by projecting
the cylindrical portion of the shell over
the firebox, which of course, allowed a
much larger steam space and much
greater disengaging surface. This modification of his original design has been
the basis on which practically all sub-
sequent locomotive boilers have been
modeled. As an example of such a
boiler, an American locomotive built in
1836 is shown on page 60.

The pressures carried at this time
were so low that it was not necessary

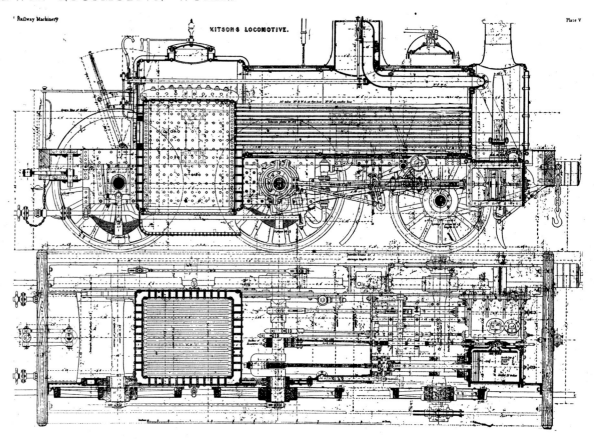

LOCOMOTIVE BUILT IN 1850

to stay the flat surfaces which appeared in the approximately rectangular fireboxes, but when the pressures
were increased and it became impossible to leave these flat surfaces unsupported, some method of preventing the
collapse of the firebox and the bulging and ultimate explosion of the shell had to be devised; the crown bars
and the staybolts therefore came into use. It is strange that no attempts were made to use some form of fire-

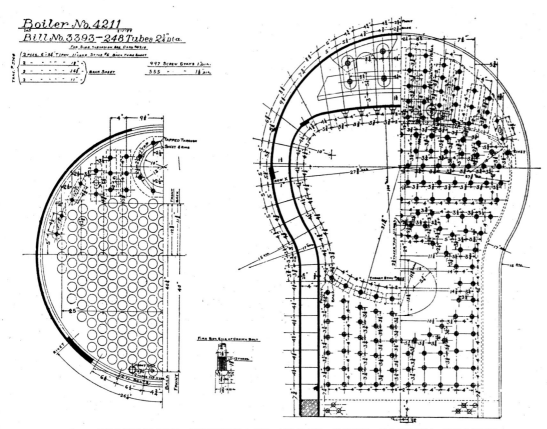

Boiler No.4211
Bill No.3393—248 Tubes 2¼"Dia.

TRANSVERSE SECTION OF BOILER WITH NARROW FIREBOX

box which would not require support, but instead of any advance being made in this direction, the developments all took the form of different methods of stays and supports for these flat and approximately flat surfaces. The boiler, therefore, seems to have followed the line of Stephenson's Multitubular Boiler in a remarkably close way. To illustrate one of the methods of support for the firebox and external shell, a drawing of a locomotive built in 1850 is shown on page 61. It is seen that longitudinal crown bars are used to support the top of the firebox and staybolts are used to support the sides.

Since the forward portion of the boiler through which the tubes run is, in all cases, cylindrical, it therefore requires no stays. For many years the crown bar boiler was used exclusively, sometimes having longitudinal and sometimes tranverse bars, but the principle of taking the pressure from the firebox crown through the medium of bolts to the crown bars and from thence to the external shell, being the same in all cases. There have also been various devices for supporting the crown bars, other than by tying them with bolts to the outer shell, such as by placing the ends of the bars on the top of the sheets forming the sides of the firebox, or on small brackets riveted to the sides of the boiler. The sides of the firebox have

BALDWIN LOCOMOTIVE WORKS

been in all cases tied to the outer shell by staybolts, the position and size of which being, of course, dependent on the thickness of the plate and on the pressure the boiler is to carry. An improvement on the crown bar boiler was the Belpaire Boiler; here the top of the boiler is made flat and parallel to the crown of the firebox, with through straight stays running perpendicular to both surfaces; the flat sides of the boiler above the crown are also supported by through stays running transversely from one side to the other. The combination of this arrangement with a wagon top in front of it is often used, the so-called wagon top being merely a gusset joining the large and small diameter in the shell to the boiler. Two of the principal disadvantages of this design are, (1) the sharp corners formed at the top of the side sheets and the ends of the top plate; and (2) the flat surfaces formed at the point where the upper portions of the side sheets run into the curved portion of the shell. The use of the Belpaire Boiler is now decreasing, its place having been taken by the Radial Stay Boiler, which has come into general use during the past few years, so that very few other types are now built. Here the firebox crown and the outer shell above it are curved, and through stays connect them, threaded at each end and

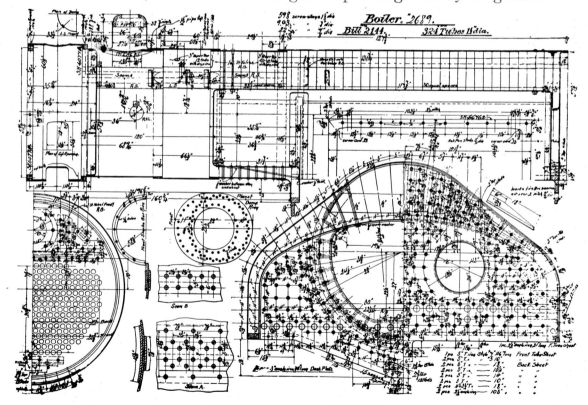

BOILER WITH WOOTTEN FIREBOX

BALDWIN LOCOMOTIVE WORKS

screwed into both sheets in the same manner as the staybolts, and so placed as to give the greatest bearing surface to the threads in each plate; the ends of the stays are riveted over on the outside after being screwed in. In designing such boilers great care must be taken to dispose the stays so as to be, as nearly as possible, normal to both surfaces, for if an entire thread does not engage in both plates there will not be sufficient bearing surface to withstand the pressure.

As regards the shape of the firebox, for many years the lower portion of the sides was narrowed in, this being of course done to to permit the firebox section of the boiler to be dropped between the frames of the loco-motive. An example of such a boiler is shown on page 62. But the Wootten Boiler, which was undoubtedly a very great advance over the type crystallized for so many years, widened out the whole firebox section and placed it over the wheels. An example of such a boiler is shown on page 63. Owing to the small space above the grates, they are prin-cipally used for burning a poor quality of soft coal. This shape of firebox is often modified; an example of such a boiler is shown herewith and on page 65.

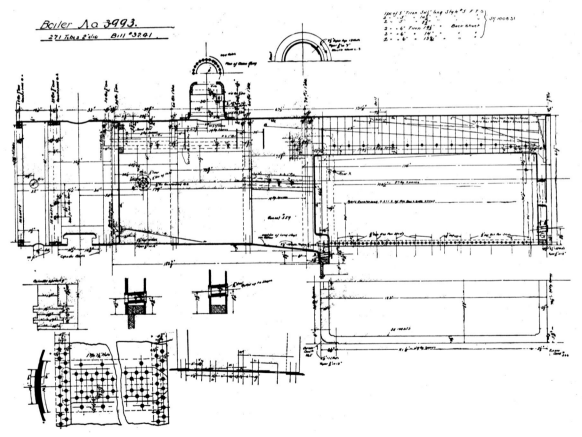

BOILER WITH WIDE FIREBOX OF THE MODIFIED WOOTTEN TYPE 64

BALDWIN LOCOMOTIVE WORKS

The methods used for staying the narrow fireboxes are equally applicable to the broad fireboxes.

It is evident that none of these designs can be carried out without the use of an enormous number of staybolts. In some of the larger boilers in use at present, anywhere from 1400 to 2000 staybolts are required for the support of the firebox and the outer shell surrounding it, and any one at all familiar with the maintenance of railway motive power, knows what a large item in the expense account is due to these staybolts. They break or leak continually, and there is no section of the locomotive causing more expense, trouble and annoyance than this

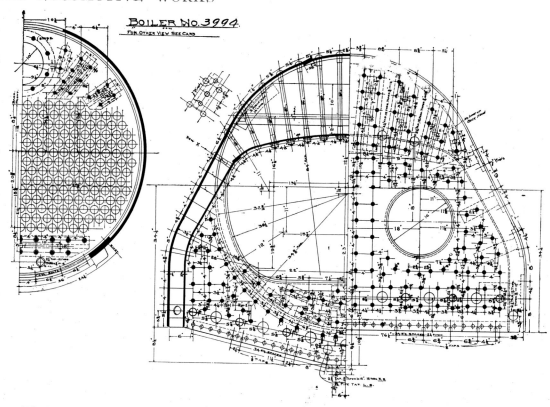

TRANSVERSE SECTION OF BOILER WITH MODIFIED WOOTTEN FIREBOX

bolted portion. When high pressures are carried, it becomes necessary to place the bolts four inches or less on centers, and they consequently fill up a large portion of the water-legs and the space above the crown sheet. When bad water is used, they become encrusted and still further prevent circulation, and the more they become encrusted, the more unequally do the firebox crown and sides expand and contract in relation to the outer shell, and therefore, the more the staybolts leak or break. As an example of the trouble occasioned by these bolts, the following quotation of a statement made by Mr. G. W. West, Superintendent of Motive Power of the New

BALDWIN LOCOMOTIVE WORKS

York, Ontario and Western R. R. at a discussion of the Wootten Boilers at the New York Railway Club, April 20, 1899, is interesting:

"We have sixty-seven engines on the Ontario and Western fitted with the wide firebox. About forty of those are nine years old, having been built in 1890. Those sixty-seven engines broke 2,703 staybolts in 1898. The average for all the engines was twenty-three and a half per engine. We have a total of one hundred and twenty-three engines, but only sixty-seven of them have the wide firebox. * * * The highest number broken in any one engine was 120. These engines all carried 160 pounds steam, except eight which are carrying 180 pounds. I have a record here of three engines for three years. Of the three engines, one had 112 broken staybolts, another 186, and the third 216 in the three years. This is a better record than we have had from our narrow firebox boilers. The total number of the bolts broken and number broken yearly was less than three per cent. of the bolts in boiler."

It was to overcome this staybolt problem that the so-called Vanderbilt Locomotive Boiler was designed, it having occurred to the author that if a cylindrical firebox could be introduced in the locomotive boiler, enormous savings in repairs would result. The allowable clearances, the position of the boiler with respect to the wheels, the allowable weights, the grates and space above them to allow for the large amount of coal that must be burned per square foot of grate, all enter very considerably into any such design. The preliminary design met with some success and proved that it was possible to satisfactorily overcome the various difficulties in the way of using a cylindrical firebox, provided a firebox of large diameter could be designed and built of sufficient strength to resist the boiler pressure. It is a well-known fact of course, that cylindrical fireboxes are in almost universal use in marine boilers. These fireboxes are corrugated in various forms to give the requisite strength, but the diameters are small, generally not over three or three and one-half feet. Upon investigating the matter, it was found that a firebox of the size contemplated in the design, namely fifty-nine inches internal diameter of three-quarter inch plate, with Morison corrugations, could be rolled and would have sufficient strength. The designer received so much encouragement from the engineers to whom the preliminary drawings were shown, that the New York Central and Hudson River R. R. decided to build a ten-wheel locomotive equipped with such a boiler. It was at this time in looking up patents relative to boiler construction, the author discovered that attempts had been made to use corrugated cylindrical fireboxes on locomotives. The Strong Boiler with two cylindrical furnaces and a combustion chamber in front, did not appear to offer the advantages of a single large flue. It was found that attempts to use a single flue had been made in Germany and had failed. These German boilers were built under the Lentz patent for boiler construction. His design is for an entirely stayless boiler having no stays or braces of any sort and the claims in his patent having relation to the disposal of front and rear course sheets as gussets, so as to avoid requiring any backhead or front tube sheet stays. That is, the central portion of the boiler has the greatest diameter and from either end of it, the sheets slope to the front tube sheet and the rear end of the boiler. The front tube sheet is entirely stayed by the flues, and there is no backhead, the course sheet terminating at the rear end of the firebox. It is evident that this design shows serious defects, especially in the lack of steam space and restricted area of disengaging surface, and to over-

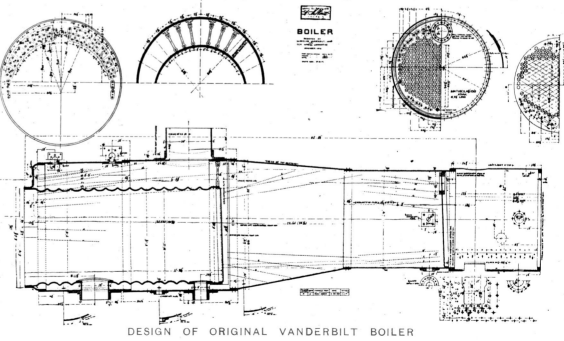

DESIGN OF ORIGINAL VANDERBILT BOILER

CORRUGATED   FIREBOX   FOR   VANDERBILT   BOILER

68

BALDWIN LOCOMOTIVE WORKS

come this, Lentz placed an auxiliary dome on the rear course, and in some designs connected it with an outside steam pipe to the main dome on the central portion of the boiler. In some of these German designs, a far more serious defect was found. In order to lower the firedoor, the firebox was bent, the axis of the front portion being parallel to the rail, and the axis of the rear portion inclined downwardly and rearwardly. To make such a firebox, it was necessary to build it up of two separate sections and to weld the joint between them. This naturally left a space of six inches or so on the top without any corrugations to assist in its support, and the firebox was weakest where it should have been strongest. The inevitable result followed the use of such a boiler; that is, the the firebox collapsed and the boiler exploded. This explosion coupled with the fact that no particularly good results were noticeable in the use of these boilers, put an end to the employment of cylindrical fireboxes in European locomotives. That no particularly good results were obtained is due, in the author's opinion, to the the simple fact that the fireboxes were too small to allow sufficient space above the grates for the proper mingling of the gases, and the small combustion chambers beyond the grates were also entirely inadequate for this purpose. Fireboxes of about the size employed in marine boilers were used, but without the combustion chambers to give space to properly consume the gases, and the result was, the combustion was unsatisfactory.

To quote from Demoulin's work on the locomotive in regard to these boilers: —

"Several inventors being anxious to improve on the ordinary expensive locomotive firebox with its numerous staybolts, have applied to the locomotive boiler the corrugated Fox type of firebox, among others, Strong, Webb and Lentz. These applications have not been crowned with great success, and cannot well be, not from the point of view of the construction itself, but of the efficiency of the boiler. The cylindrical fireboxes by their small volume and on account of the little free distance above the grate, do not lend themselves to the forced combustion of the locomotive. That is to say there is no combustion chamber and the gases are not sufficiently mingled to overcome this difficulty. Strong and later Webb placed one cylindrical firebox over a similar one; the lower

BALDWIN LOCOMOTIVE WORKS

held the grate and communicated with the upper cylinder which filled the office of combustion chamber, and carried the flue sheet.''

The American Railway clearances and limiting weights allow far larger boilers than those used on European railways, and for this reason cylindrical fireboxes of very large diameter can be used ; and in order to get sufficient space over the grates advantage can be taken of the relation of the area of a circle to its diameter.

The design of the orignal Vanderbilt Boiler is shown on page 67. It was built for a ten wheel locomotive constructed at the West Albany shops of the New York Central and Hudson River R. R., and completed in August, 1899. The total weight of the locomotive is 160,000 pounds, 113,300 of which is on the drivers and 46,700 on the truck. The cylinders are nineteen and one-half inches, by twenty-six inches, and drivers sixty-one inches ; the tube heating surface is 2165 square feet, the firebox heating surface 135 square feet, grate area thirty-four square feet. The firebox is of the Morison suspension type, with an internal diameter of fifty-nine inches, and is eleven feet, two and one-quarter inches long. An illustration of this firebox is shown on page 68. It was made by the Continental Iron Works, Greenpoint, Brooklyn, and is the largest corrugated furnace ever rolled. It was tested under an external pressure of 500 pounds per square inch before being put in the boiler, where it is carried at its front end by a row of radial sling stays and supported at its rear end by the backhead. The grates, which are not shown in the boiler drawing, run from the rear end seven feet nine inches to a bridge wall which is carried by half-round iron, resting in one of the corrugations. There is a brick arch on top of this, and the inside of the backhead is also lined with firebrick. The space in front of the brick arch is used as a combustion chamber and allows space for the proper mingling of the products of combustion, and also allows the gases to be drawn through the lower tubes.

It is seen that the axis of the rear portion of the boiler shell is inclined towards the rear downwardly. This allows a lower firedoor and also causes better draft through the lower tubes since it permits the front end of the

BALDWIN LOCOMOTIVE WORKS

grates and the brick arch to be lower. The firebox is placed eccentrically in the rear portion of the boiler; that is, its axis is inclined towards the rear downwardly with respect to the axis of the shell.

The following are the results of a road test of this locomotive on one of its first trips in regular freight service :

Road Test of New York Central and Hudson River R. R. Engine No. 947.

Mean Temperature, atmosphere	78 degrees F.
Temperature, water	71 degrees F.
Direction of wind	Unfavorable.
Velocity	Light.
Condition of rail	Good, dry.
Exhaust nozzel	Double 3¼ inch tip.
Left West Albany	9.45 A. M.
Arrived at DeWitt	6.00 P. M.
Elapsed time	8 hours, 15 minutes.
Number of slow downs	6
Number of detentions	8
Time lost by detentions	1 hour, 24 minutes.
Running time	6 hours, 51 minutes.
Distance	140 miles.
Speed, miles per hour, running time	20.4
Coal used, actual	12,390 lbs.
Water used, gallons	12,290.6
Water, pounds	107,571.6
Evaporation per pound coal	8.6
Evaporation per pound of coal from and at 212 degrees F.	10.3
Weight of train, tons 2000 lbs. each, exclusive of engine and tender	924.495 tons.
Weight of train, including engine and tender	1,054.495 tons.
Train, 60 light, 1 load, 1 caboose.	
Coal per car per mile	1.43 lbs.
Coal consumed per 100 tons, hauled one mile	9.4
Tons hauled one mile, per pound coal	10.4
Tons hauled one mile, per pound water	1.23
Quality of coal	Clearfield Bituminous.

BALDWIN LOCOMOTIVE WORKS

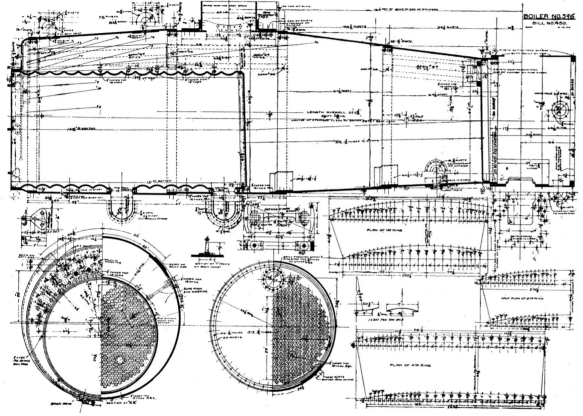

VANDERBILT BOILER FOR SCHENECTADY LOCOMOTIVES

An illustration of this locomotive is shown on page 58. It is seen that the appearance of such an engine does not materially differ from one equipped with the usual boiler, except that it is possible to look across under the boiler in the space generally occupied by the flat-sided firebox. This locomotive has been running since its delivery, in regular fast freight service on the Mohawk Division of the New York Central and Hudson River R. R. Last August, after it had been in service for one year, it was taken to the shop for the usual overhauling, and a statement of the actual cost of all material and repairs made during the first year's service, shows that the total cost of labor was $1,039.45, and materials $417.07. This includes several items not properly chargeable and which should certainly be excluded in any comparison of cost of repairs due to boilers. That is to say, one driving axle had to be replaced with a new one, and the backhead casting which was originally put in was found to be defective and had to be replaced, but in order to avoid any criticism of partiality every item of expense has been included. The total for labor and material is $1,456.52. The number of miles run was 54,650, with 9019 cars west and 7616 cars east. The small number of cars hauled is due to the fact that the engine was in fast freight service. The cost of repairs per engine mile is, therefore, 2.66 cents. This is very low in comparison with the average repairs to locomotives, especially when it is considered that this locomotive was, to a certain extent, an experiment, and no doubt had much more time devoted to it than was actually necessary, as well as to the alterations and changes, which, although small added to the expense of maintenance. The cost of repairs of all engines on the New York Central and Hudson River R. R. during the same year was 3.76 cents per engine mile, and on the Mohawk Division was 3.97 cents. This, of course, includes all the old as well as the new locomotives and cannot be used for exact comparison.

The success of this original boiler led the New York Central and Hudson River R. R. to order five more locomotives with Vanderbilt Boilers. Three of these were built at the Schenectady Locomotive Works, and the other two at the Baldwin Locomotive Works. They were ordered in November, 1899, and were delivered in May and June, 1900.

The boiler drawing of one of these Schenectady engines is shown on page 72. The criticism had been made in the original boiler that there was not a sufficient distance from the upper

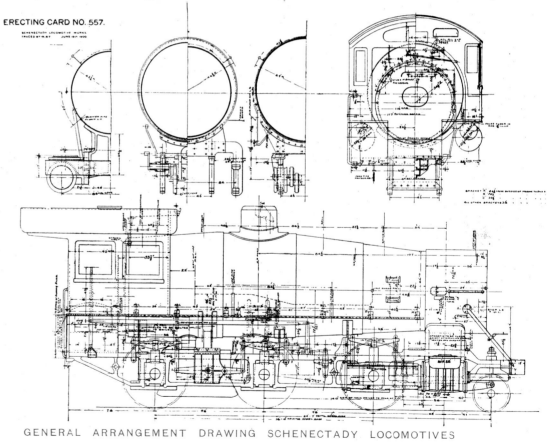

ERECTING CARD NO. 557.

GENERAL ARRANGEMENT DRAWING SCHENECTADY LOCOMOTIVES

MOGUL FREIGHT LOCOMOTIVE WITH VANDERBILT BOILER
BUILT BY THE BALDWIN LOCOMOTIVE WORKS FOR THE NEW YORK CENTRAL AND HUDSON RIVER R. R.

BALDWIN LOCOMOTIVE WORKS

gauge water level to the throttle, and therefore wet steam resulted. Although this was disputed, it was thought advisable to increase the distance between the top of the firebox at its front end and the top of the outer shell. Therefore, the maximum diameter of the boiler is very great. i. e., eighty-eight inches, and in order to economize weight as much as possible, the front and rear courses slope towards the smokebox and backhead respectively and owing to the decrease in allowable length, the straight course which was used in the original boiler, had to be omitted. This modification was introduced at the suggestion of the Baldwin Locomotive Works, and this design more nearly approaches Lentz's, but does not carry his stayless idea to nearly such an extent, as longitudinal crow-foot stays are used for the backhead. This method of reducing weight is often resorted to in the ordinary boilers of modern locomotives. There are some differences in the Schenectady and Baldwin designs. A general arrangement drawing of a Schenectady locomotive is shown on page 73. The cylinders are twenty inches by twenty-eight inches, the drivers fifty-seven inches, the weight on drivers is 145,000 pounds and on truck 25,000 pounds, a total of 170,000 pounds. The heating surface is 2732 square feet.

An illustration of one of the Baldwin engines is shown on page 74. The cylinders and wheels are of the same dimension as the Schenectady's, but the total weight is 167,500 pounds, 20,200 of which is on the truck and 147,300 on the drivers. The heating surface is 2720 square feet, with 2585 square feet in the tubes and 135 square feet in the firebox. The grate area is thirty-three square feet, with rocking grates shaking in three sections. The boilers carry 190 pounds pressure, and the fireboxes are of the same diameter, length and thickness as in the original boiler. In the Schenectady engines, a row of sling stays is introduced as in the original boiler, to assist in supporting the front end of the firebox, but they were left out of the Baldwin engines, the weight of the furnace being taken on the casting which surrounds the hole in the combustion chamber.

These five engines have been in regular service on the Hudson Division of the New York Central since their delivery; but it is, of course, impossible at present to give any statement of the cost of repairs, as sufficient

BALDWIN LOCOMOTIVE WORKS

time has not elapsed; but tests for the purpose of determinating the evaporative efficiency of the boilers in comparison with a boiler of the ordinary type, have been made. The results of two runs of the Baldwin locomotive No. 1766, equipped with Vanderbilt Boiler, in comparison with engine No. 1753 equipped with ordinary boiler, are in favor of No. 1766. No. 1753 was taken as most nearly resembling No. 1766, as it was built about the ne time by the same works, and has cylinders, wheels and wheel base of similar dimensions. The weight of the locomotive is less, being 156,200 pounds total, of which 20,700 pounds is on the truck and 135,500 pounds on the drivers. The heating surface is 2509 square feet, and the grate area is thirty square feet. The boiler carries 185 pounds steam pressure.

The engines were run from Albany to New York with no departure from the usual methods of operating, with the exception that they were turned at New York with little delay, thus making the run, as nearly as possible, a round trip. They were in fast freight service and hauled trains as nearly as could be judged of the same tractive power, and that this object was attained will be shown by the narrow limits of variations in the boiler horse-power. One car of coal was set aside for use during the tests, and the coal was weighed on and off the tenders at Albany. The water was measured by means of Thomson water meters and was checked by the use of floats in the tank. The smokebox temperatures were measured by means of a thermometer specially constructed for such work, and the smokebox vacuum was measured by means of a U tube water column.

The tables on following pages give the comparative dimensions of the two locomotives and the results of the test:

Baldwin Locomotive Works—Department of Tests

Dimensions of Engines with Vanderbilt and Standard Boilers

ITEMS	STANDARD MOGUL No. 1753	VANDERBILT BOILER No. 1766
Diameter of cylinder, inches	20.	20.
Stroke of cylinder, inches	28.	28.
Diameter of driving wheel, inches	57.	57.
Weight on truck	20,700	20,200
" " No. 1 driver	45,000	42,300
" " No. 2 "	47,000	54,200
" " No. 3 "	43,500	50,800
The largest diameter of boiler	73½ inches	88 inches
The smallest " "	66 "	67¼ "
Number of tubes	366	517
Diameter of tubes	2 inches	1¾ inches
Length of tubes	146½ "	135 "
Tube heating surface	2,323.6	2,585
Firebox heating surface	185.6	135
Total heating surface	2,509.2	2,720
Grate area	30.3	33
Steam pressure carried	185	190
Diameter of exhaust nozzle	5¼ inches	5¼ inches

Baldwin Locomotive Works—Department of Tests

Comparative Test of Vanderbilt and Standard Boilers

	ENGINE No. 1753 STANDARD MOGUL	ENGINE No. 1766 VANDERBILT BOILER	ENGINE No. 1766 VANDERBILT BOILER
FUEL—POUNDS			
Consumed on run	28,870.	31,560.	30,400.
" per hour, running time	2,392.5	2,431.4	2,403.2
WATER—POUNDS			
Total water evaporated on run	202,550.	224,890.	233,088.
Water consumed, per hour	16,785.9	17,325.9	18,425.9
Equivalent weight evaporated from and at 212° on run	246,908.45	274,590.69	285,066.87
" " " " per hour	20,462.01	21,154.9	22,527.
Boiler, horse-power	593.27	613.36	653.14
ECONOMIC EVAPORATION—POUNDS			
Water evaporated, per pound of coal, on run	7.02	7.12	7.66
Equivalent evaporation from and at 212°	8.56	8.69	9.01
RATE OF COMBUSTION—POUNDS			
Coal burned, per square foot, grate surface	951.86	956.36	921.21
" " " " per hour	78.88	73.68	72.82
RATE OF EVAPORATION—POUNDS			
Water evaporated, per square foot heating surface	80.72	82.64	85.69
" " " " per hour	6.69	6.37	6.77
" from and at 212°, evaporated, per square foot heating surface, per hour	8.15	7.77	7.93
Square feet heating surface, per horse-power	4.21	4.43	4.16

	SOUTH	NORTH	SOUTH	NORTH	SOUTH	NORTH
TIME DATA						
Running time	5 hrs. 53 min.	6 hrs. 11 min.	7 hrs. 16 min.	5 hrs. 43 min.	7 hrs. 2 min.	5 hrs. 37 min.
Total time on road	8 hrs. 05 min.	10 hrs. 35 min.	17 hrs. 50 min.	6 hrs. 32 min.	9 hrs. 52 min.	9 hrs. 09 min.
Delays	2 hrs. 12 min.	4 hrs. 24 min.	10 hrs. 34 min.	49 min.	2 hrs. 50 min.	3 hrs. 22 min.
TRAIN DATA						
Date	Nov. 14, 1900	Nov. 15, 1900	Nov. 16, 1900	Nov. 17, 1900	Nov. 18, 1900	Nov. 19, 1900
Train	2 Ex.	N. B. 3	6 Ex.	N. R. 1	5 Ex.	2 Ex.
Number of loaded cars	39	40	40	40	39	
" empty "	1				1	49
GENERAL DATA						
Average steam pressure, when developing power, pounds	173.	177.	180.	194.7	194.	192.8
" smoke-box vacuum, " " " inches water	4.5	5.3	5.2	4.6	5.8	6.7
" " temperature, degrees	721.	734.	741.	785.	751.	763.
" temperature of feed water, degrees	50.	50.	50.	50.	50.	50.
" speed under headway	25.1	26.3	23.	28.6	22.9	27.7

BALDWIN LOCOMOTIVE WORKS

It is seen that the running conditions were very unfavorable to engine No. 1766 in that the delays were more frequent and of greater duration than those of engine No 1753 and while the south-bound speed of the latter engine was greater than that of the former, it was due to the numerous slow orders. Between the delay the engine was worked so hard that the demands upon the boiler were variable and fluctuating, the most unfavorable conditions under which a test can be conducted. Notwithstanding this, the cylindrical firebox engine consumed 3.68 pounds of coal per horsepower hour on one test and 3.96 pounds on the other, an average of 3.82 pounds. Whereas, the standard boiler consumed 4.04 pounds, showing a saving of 5.4 per cent. in favor of the Vanderbilt Boiler. This saving is in a measure, due to the larger grate area and heating surface of engine No. 1766. These larger surfaces should have lowered the smokebox vacuums and temperatures, but such results were not obtained because of the fact previously stated, that the engine with the newer type of boiler was worked harder than the one

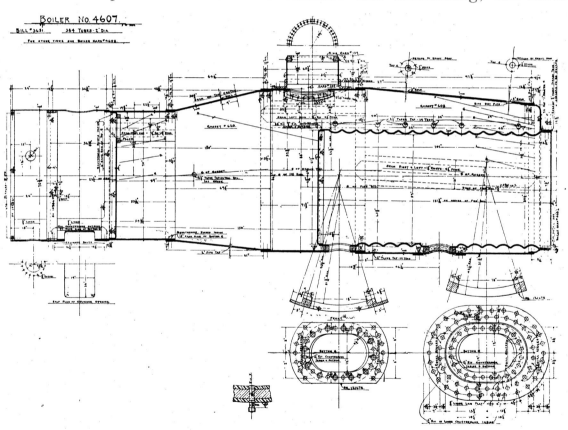

BOILER OF LOCOMOTIVE FOR THE UNION PACIFIC R. R
BUILT BY THE BALDWIN LOCOMOTIVE WORKS

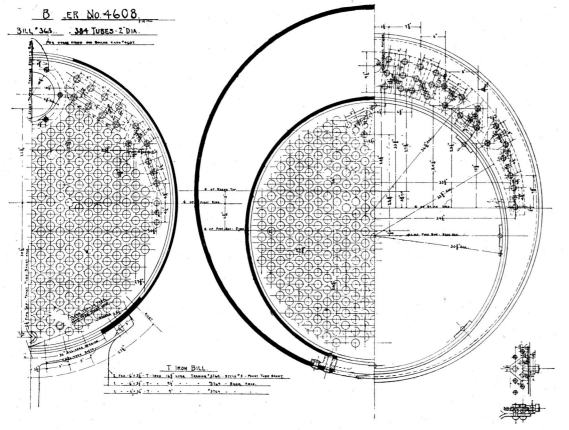

TRANSVERSE SECTION OF BOILER, UNION PACIFIC LOCOMOTIVE

with the old; but that the cylindrical firebox form of boiler is more efficient than the ordinary, is shown by a comparison of the economic evaporation of both boilers. The average result in the two tests of the Vanderbilt Boiler is 8.85 pounds of water per pound of coal, from and at 212 degrees F., and in the standard boiler 8.56 pounds; a gain of .29 pounds for the cylindrical firebox type. Both boilers developed one horsepower for each four and one-quarter square feet of heating surface, and steamed freely, carrying their water well.

Soon after the order for these five engines, the Union Pacific Railway ordered two Consolidation locomotives with Vanderbilt Boilers to be built by the Baldwin Locomotive Works, and shortly afterwards the Baltimore and Ohio R. R. placed a similar order. The Union Pacific's were delivered last July, and the Baltimore and Ohio's in August, and since their delivery have been running in regular service on their respective roads.

The boiler design of one of the Union Pacific locomotives is shown on page 79. It is seen that it was

BALDWIN LOCOMOTIVE WORKS

possible to go back more nearly to the original design, that is to say, the increased length allowable in the boilers of a Consolidation or ten-wheel engine permits the use of a front course of small diameter, and does away with the necessity of sloping the front course into the smokebox. These boilers have a heating surface 2629 square feet, 2494 square feet being in the tubes and 135 square feet in the firebox. It is secured by 384 two-inch tubes, twelve feet six inches long; the fireboxes are of the same dimensions as in the locomotives previously described. The boilers carry 190 pounds pressure. A transverse section of one of these boilers is shown on page 80, and the erecting card of one of these engines is shown herewith. The total weight is about 196,000 pounds, with 174,000 pounds

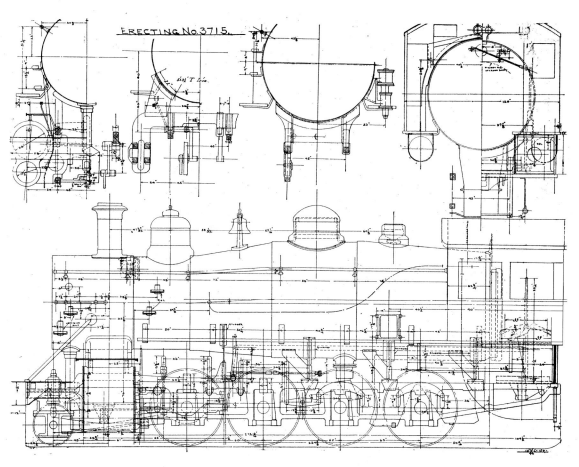

ERECTING CARD OF LOCOMOTIVE FOR THE UNION PACIFIC R. R.
BUILT BY THE BALDWIN LOCOMOTIVE WORKS

on the drivers and 22,000 pounds on the truck. The cylinders are Vauclain Compound fifteen and one-half inches and twenty-six inches by thirty inches; the drivers are fifty-seven inches in diameter, with a total wheel base of twenty-three feet eleven inches, of which fifteen feet three inches is rigid. An

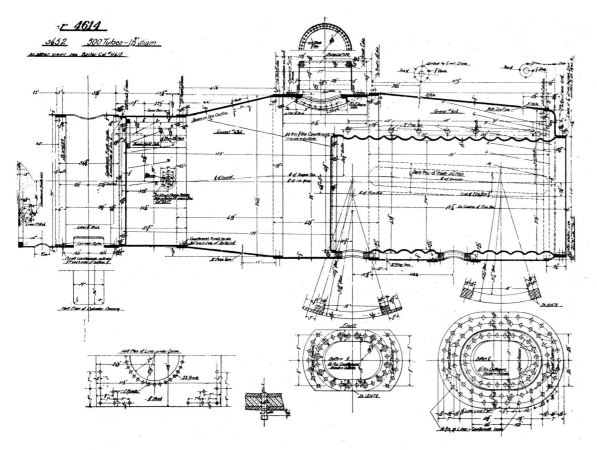

BOILER OF LOCOMOTIVE FOR THE BALTIMORE AND OHIO R. R.
BUILT BY THE BALDWIN LOCOMOTIVE WORKS

illustration of one of these engines is shown on page 83.

One of the objects in building the Union Pacific engines was to determine whether this form of boiler was not more satisfactory for bad water than the staybolt form. Therefore, as soon as they were delivered they were placed in regular service on the division on which the worst water is encountered. Up to the present time all reports as to their work and operation have been most satisfactory.

The two Consolidations built for the Baltimore and Ohio are very similar to the Union Pacific's. The boiler drawing of one is shown herewith. Instead of the two-inch tubes, one and three-quarter inch tubes are used owing to the decreased length of boiler There are 500 of them, eleven feet six inches long, giving a heating surface of 2615 square feet. Including 135 square feet in the firebox, the total is 2750 square feet. A transverse section is shown on page 84.

The erecting card of one of these engines is shown on page 85. The cylinders are Vauclain Compound,

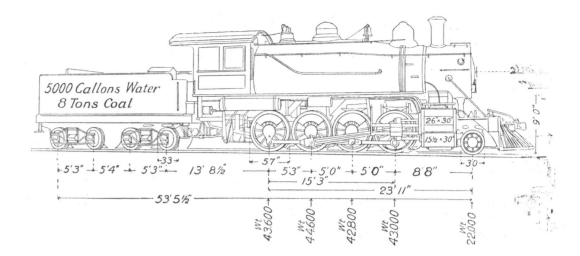

5000 Gallons Water
8 Tons Coal

CONSOLIDATION LOCOMOTIVE WITH VANDERBILT BOILER
BUILT BY THE BALDWIN LOCOMOTIVE WORKS FOR THE UNION PACIFIC R. R.

BALDWIN LOCOMOTIVE WORKS

fifteen and one-h f inches and twenty-six inches by thirty inches. The weight on drivers is about 170,800 pounds and on truck, 23,100 pounds, a total of 193,900. The wheels are fifty-four inches diameter, with a total wheel base of twenty-three feet eight inches, fifteen feet four inches of which is rigid.

An illustration of one of these locomotives is shown on page 87. As yet no tests for evaporative efficiency have been made on these last four engines, and it is of course impossible to determine the relative cost of repairs. All indications, however, point to a material decrease in the cost of maintenance in such a form of boiler, for there seems to be but little to balance the present cost of repairs and renewals due to the staybolts. It is also apparent that the firebox can be renewed in much less time and at much less cost than the stayed type. There should also be less difficulty in keeping the flues tight by reason of the protection afforded by the combustion chamber and because of the uniformity in the expansion of the firebox. There is, of course, more firebrick to be renewed, but this is a comparatively small matter. As regards the weight of a

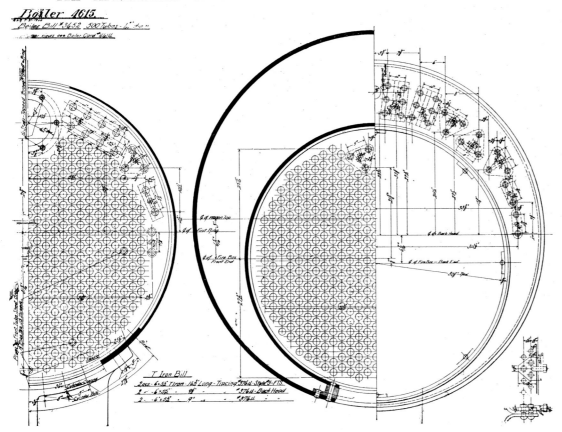

TRANSVERSE SECTION OF BOILER FOR BALTIMORE AND OHIO LOCOMOTIVE

BALDWIN LOCOMOTIVE WORKS

cylindrical firebox boiler compared with a narrow firebox boiler, a comparison of the total weight of engine per square foot of heating surface, shows in the case of the two Mogul engines previously compared, that the Vanderbilt Boiler engine is slightly lighter than the standard.

The absence of stays largely removes localized stresses in the firebox, the cause of the cracking of the firebox sheets. The rear end of the boiler is also far more accessible for cleaning and removal of scale. In the original boiler only a very slight coating, less than one thirty-second of an inch of scale was discovered on the outside of the firebox after it had been in service for one year; whereas, in the usual type, from one-sixteenth to one-eighth of an inch is always taken off after similar service.

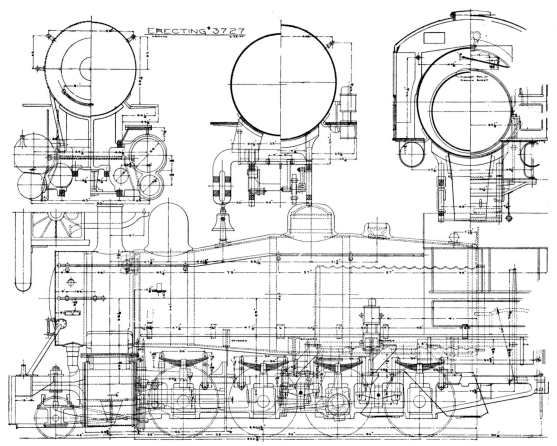

ERECTING CARD OF LOCOMOTIVE FOR THE BALTIMORE AND OHIO R. R.
BUILT BY THE BALDWIN LOCOMOTIVE WORKS

As regards the relative strength of flat stayed surfaces and cylindrical flues, the results obtained in a test made by the Baldwin Locomotive Works in 1897, to determine the effect of repeatedly applied loads to the ordinary flat stayed surfaces, may be compared with those obtained from experiments with Morison furnaces in

BALDWIN LOCOMOTIVE WORKS

1891. In the Baldwin test, pressure was applied to a drum eighteen inches in diameter, four inches deep, made of three-eighths inch steel plate with one-inch iron staybolts placed four inches on centers. Repeated applications of a pressure of 100 pounds were made for twenty-four hours, the pressure being held by an automatic apparatus for an interval of six seconds, and at the rate of 350 applications per hour. A pressure of 200 pounds was similarly applied for twenty-four hours, after which the pressures were increased by increments of fifty pounds. After the test under 350 pounds had been conducted for thirteen hours, the staybolts started to leak; this continued for the remaining eleven hours. After a pressure of 400 pounds had been applied for four hours, the heads began to crack and continued to crack for twenty hours. At 450 pounds pressure for eight hours, a complete rupture occurred by the head stripping from one of the center staybolts. The drum contained twelve bolts, and at the end of the twenty-four hours under 450 pounds, all of the heads were damaged to a greater or less extent.

From experiments conducted under the Bureau of Steam Engineering, United States Navy in 1879, it was determined that the bolts driven with spherical heads were from twenty-three per cent. to thirty-six per cent. stronger than those driven with the usual conical head. Slightly more metal was used for the spherical heads, though how much is not stated. This would indicate that after a few years' service there would be a decided decrease in the strength of a flat stayed surface from the fact that the rivet head would be gradually diminished in size from the repeated hammering incident to keeping the end tight.

The Baldwin tests indicate that a flat sided firebox three-eighths inch thick with one-inch staybolts spaced four inches on centers, which is a fair average for modern firebox dimensions, would rupture at a pressure of 450 pounds, after a certain definite number of applications of gradually increasing pressures.

From Morison's experiments on his furnaces at the Leeds Forge in 1891, he found that a firebox nine-sixteenths inch thick collapsed under a continuous pressure of 1340 pounds. The formula deduced

CONSOLIDATION LOCOMOTIVE WITH VANDERBILT BOILER
BUILT BY THE BALDWIN LOCOMOTIVE WORKS FOR THE BALTIMORE AND OHIO R. R.

BALDWIN LOCOMOTIVE WORKS

from his experiments shows that under repetitions of stress, the strength in the case of cylindrical flues depends not on the ultimate strength of the material, but upon the modulus of elasticity. This, unlike the ultimate strength, is not diminished by repeated applications of stresses. The formula derived from experiments gives a factor of safety of 5.96 for a Morison furnace three-quarter inches thick and fifty-nine inches internal diameter.

Comparing this result with that obtained from the Baldwin experiments, we find that a firebox such as has been applied to the locomotive boilers previously described is over twice as strong under repeated stresses as the ordinary flat stayed type.

As regards the other forms of corrugated fireboxes, a comparison of the various types is interesting. Taking the Morison as 100 per cent.: the relative strength per square inch of the Fox section is 91.82 per cent.; for the Purves, 89.55 per cent.; for the Adamson, 75.86 per cent.; for the Holmes, 72.93 per cent.; and 67.51 per cent. for the Farnley Spiral Flue.

A proper estimate of the advantages of the Vanderbilt boiler over those of the present type can only be appreciated after the first two years of service. Comparisons made then will, in the author's opinion, justify the advocated departure from the present type of screw-bolted fireboxes in use almost exclusively since the advent of the locomotive.

Compound Consolidation Locomotive with wide firebox on the Lehigh Valley Railroad near Coxton Yard. Coupled to a full train weighing one thousand tons and ready to ascend the grade of sixty feet per mile.

Full particulars are given of a locomotive of this class in Record of Recent Construction No. 10, pages 2 and 3.

BALDWIN LOCOMOTIVE WORKS

Compound Ten-Wheel Locomotive

Class 10 $\frac{24}{44}$-D-54

for the

Gauge 4' 8½"

Baltimore and Ohio Railroad Company

GENERAL DIMENSIONS

CYLINDERS

Diameter (High Pressure) . .	15"
" (Low Pressure)	25"
Stroke	28"
Valve . . . . Balanced Piston	

BOILER

Diameter	60"
Thickness of Sheets . ⅝" and 11⁄16"	
Working Pressure . . .	200 lbs.
Fuel	Coke

FIREBOX

Material	Steel
Length	120⅜"
Width	41⅛"
Depth (front) . . .	72⅝"
" (back) . . .	60⅛"
Thickness of Sheets, Sides .	5⁄16"
" " " Back .	⅜"
" " " Crown .	⅜"
" " " Tube .	½"

TUBES

Material . . . . . Iron	
Number	231
Diameter	2¼"
Length	14' 11"

HEATING SURFACE

Firebox . . .	181 sq. ft.
Tubes . . .	2,018 sq. ft.
Total	2,199 sq. ft.
Grate Area . . .	34.4 sq. ft.

DRIVING WHEELS

Diameter Outside . . .	78"
" of Center . .	72"
Journals	9" x 12"

ENGINE TRUCK WHEELS

Diameter	33"
Journals . . .	5½" x 10"

WHEEL BASE

Driving	13' 6"
Total Engine . . .	25' 7"
Total Engine and Tender .	52' 10"

WEIGHT

On Drivers . . .	114,800 lbs.
On Truck . . .	41,300 lbs.
Total Engine . . .	156,100 lbs.
Total Engine and Tender	256,000 lbs.

TENDER

Diameter of Wheels . .	36"
Journals	5" x 9"
Tank Capacity . .	5000 gals.

SERVICE

Passenger.

BALDWIN LOCOMOTIVE WORKS

Compound Ten-Wheel Locomotive

Class 10 $\frac{28}{50}$-D-1

for the

Gauge 4' 8½"

Lehigh Valley Railroad Company

GENERAL DIMENSIONS

CYLINDERS

Diameter (High Pressure) .	17"
" (Low Pressure) . .	28"
Stroke	26"
Valve . . .	Balanced Piston

BOILER

Diameter	64"
Thickness of Sheets . ⅝" and 11⁄16"
Working Pressure . .	200 lbs.
Fuel . . .	Anthracite Coal

FIREBOX

Material	Steel
Length	114"
Width	90"
Depth, Front . . .	52⅜"
" Back . . .	50⅜"
Thickness of Sheets, Sides .	⅜"
" " " Back . .	⅜"
" " " Crown .	⅜"
" " " Tube .	½"

TUBES

Material	Iron
Number	325
Diameter	2"
Length	15' 0"

HEATING SURFACE

Firebox . . .	171.71 sq. ft.
Tubes . . .	2536.59 sq. ft.
Total . . .	2708.3 sq. ft.
Grate Area . .	71.25 sq. ft.

DRIVING WHEELS

Diameter Outside . . .	72"
" of Center . .	66"
Journals, Main . . .	10" x 12"
" Others . .	9" x 12"

ENGINE TRUCK WHEELS

Diameter	33"
Journals	6" x 12"

WHEEL BASE

Driving	13' 0"
Total Engine . . .	25' 3½"
Total Engine and Tender .	52' 6⅞"

WEIGHT

On Drivers . . .	138,348 lbs.
On Truck . . .	53,410 lbs.
Total Engine . . .	191,758 lbs.
Total Engine and Tender .	282,000 lbs.

TENDER

Diameter of Wheels . .	36"
Journals . . .	4¼" x 8"
Tank Capacity . .	4,500 gals.

SERVICE

Passenger.

BALDWIN LOCOMOTIVE WORKS

Compound Mogul Locomotive

Class 8 $\frac{25}{46}$-D-24

for the

Gauge 4' 8½"

Atchison, Topeka & Santa Fe Railway Company

GENERAL DIMENSIONS

CYLINDERS

Diameter (High Pressure) . .	15½"
" (Low Pressure) .	26"
Stroke	28"
Valve . . .	Balanced Piston

BOILER

Diameter	68"
Thicknes of Sheets	¾"
Working Pressure . .	200 lbs.
Fuel . .	Bad Bituminous Coal

FIREBOX

Material	Steel
Length	100¾"
Width	71¼"
Depth, Front . . .	69"
" Back . . .	54½"
Thickness of Sheets, Sides .	⅜"
" " " Back .	⅜"
" " " Crown .	⅜"
" " " Tube .	½"

TUBES

Material	Iron
Number . .	350
Diameter .	2"
Length . . .	13' 5"

HEATING SURFACE

Firebox . . .	155.4 sq. ft.
Tubes . . .	2443.4 sq. ft.
Total . . .	2598.8 sq. ft.
Grate Area . . .	48 sq. ft.

DRIVING WHEELS

Diameter Outside . . .	62"
" of Center . .	56"
Journals . . .	9" x 12"

ENGINE TRUCK WHEELS

Diameter	30"
Journals . . .	6" x 10"

WHEEL BASE

Driving . . .	16' 0"
Total Engine . .	24' 8"
Total Engine and Tender .	51' 4"

WEIGHT

On Drivers . . .	135,000 lbs.
On Truck . . .	29,000 lbs.
Total Engine .	164,000 lbs.
Total Engine and Tender	264,000 lbs.

TENDER

Diameter of Wheels . .	33"
Journals . . .	5" x 9"
Tank Capacity . .	5,000 gals.

SERVICE

Freight.

BALDWIN LOCOMOTIVE WORKS

Mogul Locomotive

for the

Atchison, Topeka & Santa Fe Railway Company

Class 8-34-D-134

Gauge 4' 8½"

GENERAL DIMENSIONS

CYLINDERS

Diameter	20"
Stroke	28"
Valve	Balanced Piston

BOILER

Diameter	68"
Thickness of Sheets	¾"
Working Pressure	200 lbs.
Fuel	Bad Bituminous Coal

FIREBOX

Material	Steel
Length	100¾"
Width	71¼"
Depth (front)	69"
" (back)	54½"
Thickness of Sheets, Sides	⅜"
" " " Back	⅜"
" " " Crown	⅜"
" " " Tube	½"

TUBES

Material	Iron
Number	350
Diameter	2"
Length	13' 5"

HEATING SURFACE

Firebox	155.4 sq. ft.
Tubes	2443.4 sq. ft.
Total	2598.8 sq. ft.
Grate Area	48 sq. ft.

DRIVING WHEELS

Diameter Outside	62"
" of Center	56"
Journals	9" x 12"

ENGINE TRUCK WHEELS

Diameter	30"
Journals	6" x 10"

WHEEL BASE

Driving	16' 0"
Total Engine	24' 8"
Total Engine and Tender	51' 4"

WEIGHT

On Drivers	132,000 lbs.
On Truck	25,100 lbs.
Total Engine	157,100 lbs.
Total Engine and Tender	257,100 lbs.

TENDER

Diameter of Wheels	33"
Journals	5" x 9"
Tank Capacity	5,000 gals.

SERVICE

Freight.

BALDWIN LOCOMOTIVE WORKS

Compound Consolidation Locomotive

Class 10 $\frac{28}{50}$-E-80

for the

Gauge 4′ 8½″

Rio Grande Western Railway Company

GENERAL DIMENSIONS

CYLINDERS

Diameter (High Pressure) . .	17″
" (Low Pressure) . .	28″
Stroke	30″
Valve . . .	Balanced Piston

BOILER

Diameter	76″
Thickness of Sheets . .	¾″
Working Pressure . .	200 lbs.
Fuel . . .	Soft Coal

FIREBOX

Material	Steel
Length	121¹¹⁄₁₆″
Width	41¾″
Depth, Front . . .	75″
" Back . . .	71½″
Thickness of Sheets, Sides .	⁵⁄₁₆″
" " " Back	⁵⁄₁₆″
" " " Crown	⅜″
" " " Tube	½″

TUBES

Material	Iron
Number	387
Diameter	2″
Length	15′ 6″

HEATING SURFACE

Firebox . . .	206.4 sq. ft.
Tubes . . .	3123.8 sq. ft.
Total . . .	3330.2 sq. ft.
Grate Area . .	35 sq. ft.

DRIVING WHEELS

Diameter, Outside . . .	57″
" of Center . .	50″
Journals . .	9″ x 13″

ENGINE TRUCK WHEELS

Diameter	30″
Journals . . .	5½″ x 12″

WHEEL BASE

Driving	16′ 3″
Total Engine . . .	25′ 1″
Total Engine and Tender .	53′ 5″

WEIGHT

On Drivers . . .	177,160 lbs.
On Truck . . .	21,500 lbs.
Total Engine . . .	198,660 lbs.
Total Engine and Tender .	310,000 lbs.

TENDER

Diameter of Wheels . . .	33″
Journals	5″ x 9″
Tank Capacity . .	6,000 gals.

SERVICE

Freight.

BALDWIN LOCOMOTIVE WORKS

Compound Consolidation Locomotive

Class 10 $\frac{28}{50}$-E-75

for the

Gauge 4' 8½"

Colorado Midland Railway Company

GENERAL DIMENSIONS

CYLINDERS

Diameter (High Pressure)	17"
" (Low Pressure)	28"
Stroke	30"
Valve	Balanced Piston

BOILER

Diameter	74"
Thickness of Sheets	¾"
Working Pressure	200 lbs.
Fuel	Soft Coal

FIREBOX

Material	Steel
Length	120⅛"
Width	42"
Depth, Front	65½"
" Back	61"
Thickness of Sheets, Sides	5/16"
" " " Back	3/8"
" " " Crown	3/8"
" " " Tube	½"

TUBES

Material	Iron
Number	337
Diameter	2"
Length	14' 0"

HEATING SURFACE

Firebox	172 2 sq. ft.
Tubes	2453.7 sq. ft.
Total	2625.9 sq. ft.
Grate Area	35 sq. ft.

DRIVING WHEELS

Diameter Outside	60"
Diameter of Center	52"
Journals	9" x 11"

ENGINE TRUCK WHEELS

Diameter	30"
Journals	6" x 10"

WHEEL BASE

Driving	15' 9"
Total Engine	24' 4"
Total Engine and Tender	53' 2"

WEIGHT

On Drivers	157,500 lbs.
On Truck	22,500 lbs.
Total Engine	180,000 lbs.
Total Engine and Tender	300,000 lbs.

TENDER

Diameter of Wheels	33"
Journals	5½" x 10"
Tank Capacity	6,000 gals

SERVICE

Passenger.

Grades four per cent. combined with curves of sixteen degrees.

BALDWIN LOCOMOTIVE WORKS

Prairie Type Locomotive

for the

Class 10-34 ¼-D-1

Gauge 4' 8½"

Chicago, Burlington & Quincy Railroad Company

GENERAL DIMENSIONS

CYLINDERS

Diameter	20"
Stroke	24"
Valve	Balanced Piston

BOILER

Diameter	65¼"
Thickness of Sheets	11⁄16"
Working Pressure	200 lbs.
Fuel	Soft Coal

FIREBOX

Material	Steel
Length	84"
Width	72"
Depth, Front	66¼"
" Back	63¼"
Thickness of Sheets, Sides	⅜"
" " " Back	⅜"
" " " Crown	⅜"
" " " Tube	½"

TUBES

Material	Iron
Number	272
Diameter	2¼"
Length	17' 1 11⁄16"

HEATING SURFACE

Firebox	155.8 sq. ft.
Tubes	2732.7 sq. ft.
Total	2888.5 sq. ft.
Grate Area	42 sq. ft.

DRIVING WHEELS

Diameter Outside	64"
" of Center	56"
Journals	9" x 10"

ENGINE TRUCK WHEELS

Diameter	37¼"
Journals	5½" x 9"

TRAILING WHEELS

Diameter	37"
Journals	6" x 10"

WHEEL BASE

Driving	12' 1"
Total Engine	28' 0"
Total Engine and Tender	54' 5½"

WEIGHT

On Drivers	129,575 lbs.
On Truck	14,200 lbs.
On Trailing Wheels	24,700 lbs.
Total Engine	168,475 lbs.
Total Engine and Tender	288,000 lbs.

TENDER

Diameter of Wheels	36"
Journals	5" x 9"
Tank Capacity	6,000 gals.

SERVICE

Freight.

BALDWIN LOCOMOTIVE WORKS

Six-Coupled Double Ender Locomotive

Class 12-22¼-D-5

for the

Gauge 3′ 6″

Government Railways of New Zealand

GENERAL DIMENSIONS

CYLINDERS

Diameter	14″
Stroke	20″
Valve	Balanced

BOILER

Diameter	50″
Thickness of Sheets	9/16″
Working Pressure	200 lbs.
Fuel	Coal

FIREBOX

Material	Steel
Length	83 5/16″
Width	29 7/8″
Depth, Front	49″
" Back	38¾″
Thickness of Sheets, Sides	5/16″
" " " Back	5/16″
" " " Crown	3/8″
" " " Tube	½″

TUBES

Material	Steel
Number	177
Diameter	1¾″
Length	9′ 5″

HEATING SURFACE

Firebox	89 sq. ft.
Tubes	755.3 sq. ft.
Total	844.3 sq. ft.
Grate Area	17.3 sq. ft.

DRIVING WHEELS

Diameter Outside	39¾″
" of Center	34″
Journals	6″ x 7″

ENGINE TRUCK WHEELS (Front)

Diameter	24¾″
Journals	4¼″ x 7½″

ENGINE TRUCK WHEELS (Back)

Diameter	24¾″
Journals	3½″ x 6″

WHEEL BASE

Driving	10′ 0″
Total	27′ 7″

WEIGHT

On Drivers	66,560 lbs.
On Truck, Front	12,100 lbs.
" " Back	18,600 lbs.
Total	97,260 lbs.

TANK

Tank Capacity	1,078 gals.

SERVICE

Passenger.

Grades	1 ft. in 33 ft.
Curves	5 chains radius

BALDWIN LOCOMOTIVE WORKS

Ten-Wheel Locomotive

for the

Government Railways of New Zealand

Class 10-26-D-161

Gauge 3' 6"

GENERAL DIMENSIONS

CYLINDERS

Diameter	16"
Stroke	20"
Valve	Balanced Piston

BOILER

Diameter	52"
Thickness of Sheets	½"
Working Pressure	200 lbs.
Fuel	Soft Coal

FIREBOX

Material	Steel
Length	76³⁄₁₆"
Width	30⅜"
Depth, Front	56½"
" Back	46½"
Thickness of Sheets, Sides	5⁄16"
" " " Back	5⁄16"
" " " Crown	⅜"
" " " Tube	½"

TUBES

Material	Steel
Number	177
Diameter	2"
Length	13' 4½"

HEATING SURFACE

Firebox	89.7 sq. ft.
Tubes	1230.8 sq. ft.
Total	1320.5 sq. ft.
Grate Area	16 sq. ft.

DRIVING WHEELS

Diameter Outside	49"
" of Center	44"
Journals	6½" x 7"

ENGINE TRUCK WHEELS

Diameter	26"
Journals	4¼" x 7½"

WHEEL BASE

Driving	10' 0"
Total Engine	20' 11"
Total Engine and Tender	42' 7"

WEIGHT

On Drivers	63,580 lbs.
On Truck	20,462 lbs.
Total Engine	84,042 lbs.
Total Engine and Tender	125,000 lbs.

TENDER

Diameter of Wheels	28"
Journals	3¾" x 7"
Tank Capacity	2,000 gals.

SERVICE

Passenger and Freight.

Curves 7½ chains radius, Bogies to pass sidings 5 chains radius.

BALDWIN LOCOMOTIVE WORKS

Compound Six-Coupled Double Ender Locomotive

for the

Government Railways of Western Australia

Class 12 $\frac{18}{34}$ ¼-D-1

Gauge 3' 6''

GENERAL DIMENSIONS

CYLINDERS

Diameter, (High Pressure) .	12''
" (Low Pressure) . .	20''
Stroke	22''
Valve . . .	Balanced Piston

BOILER

Diameter	56''
Thickness of Sheets . .	9/16''
Working Pressure . .	200 lbs.
Fuel	Lignite Coal

FIREBOX

Material	Copper
Length	102''
Width	29''
Depth, Front . . .	59''
" Back . . .	51¾''
Thickness of Sheets, Sides	½''
" " " Back . .	½''
" " " Crown .	⅝''
" " " Tube ¾'' and ½''	

TUBES

Material	Copper
Number	259
Diameter	1¾''
Length	12' 7''

HEATING SURFACE

Firebox	121 sq. ft.
Tubes	1478 sq. ft.
Total	1599 sq. ft.
Grate Area . . .	20.5 sq. ft.

DRIVING WHEELS

Diameter Outside . . .	54''
" of Center . .	48''
Journals	7'' x 8''

ENGINE TRUCK WHEELS (Front)

Diameter	26''
Journals . . .	4¼'' x 7½''

ENGINE TRUCK WHEELS (Back)

Diameter	26''
Journals . . .	4¼'' x 7½''

WHEEL BASE

Driving	11' 6''
Total Engine . . .	26' 3''
Total Engine and Tender .	49' 2½''

WEIGHT

On Drivers . . .	71,020 lbs.
On Truck, Front . .	24,400 lbs.
" " Back . .	7,800 lbs.
Total Engine . .	103,220 lbs.
Total Engine and Tender	163,000 lbs.

TENDER

Diameter of Wheels . .	33''
Journals . . .	4¼'' x 8''
Tank Capacity . .	3,000 gals.

SERVICE

Passenger and Freight.

BALDWIN LOCOMOTIVE WORKS

Eight-Coupled Locomotive

for the

Rio Tinto Company, Limited (Spain)

Class 8-26-E 5

Gauge 3′ 6″

GENERAL DIMENSIONS

CYLINDERS

Diameter	16″
Stroke	22″
Valve	Balanced

BOILER

Diameter	48″
Thickness of Sheets . .	½″ and 7/16″
Working Pressure . .	180 lbs.
Fuel	Coal

FIREBOX

Material	Copper
Length	77″
Width	29½″
Depth, Front . . .	44¼″
" Back . .	42¾″
Thickness of Sheets, Sides .	½″
" " " Back . .	½″
" " " Crown .	½″
"   "   " Tube ¾″ and ½″

TUBES

Material	Brass
Number	158
Diameter	1¾″
Length	10′ 11″

HEATING SURFACE

Firebox . . .	82 sq. ft.
Tubes . . .	782 sq. ft.
Total . . .	864 sq. ft.
Grate Area . . .	16 sq. ft.

DRIVING WHEELS

Diameter Outside . . .	42″
Diameter of Center . .	36″
Journals . . .	6½″ x 7″

WHEEL BASE

Driving	11′ 4″
Total	11′ 4″

WEIGHT

On Drivers . . .	89,395 lbs.
Total . . .	89,395 lbs.

TANK

Tank Capacity . .	1,200 gals.

SERVICE

Freight.

BALDWIN LOCOMOTIVE WORKS

Six-Coupled Double Ender Locomotive

for the

Zululand Railway (Natal)

Class 10-18¼-D-6

Gauge 3′ 6″

GENERAL DIMENSIONS

CYLINDERS

Diameter	12″
Stroke	18″
Valve	Balanced

BOILER

Diameter	36″
Thickness of Sheets	⅜″
Working Pressure	160 lbs.
Fuel	Soft Coal

FIREBOX

Material	Steel
Length	60″
Width	30⅜″
Depth, Front	37½″
" Back	30¼″
Thickness of Sheets, Sides	5⁄16″
" " " Back	5⁄16″
" " " Crown	⅜″
" " " Tube	7⁄16″

TUBES

Material	Iron
Number	84
Diameter	1¾″
Length	8′ 6″

HEATING SURFACE

Firebox	34.4 sq. ft.
Tubes	323.4 sq. ft.
Total	357.8 sq. ft.
Grate Area	12.6 sq. ft.

DRIVING WHEELS

Diameter Outside	42″
" of Center	36″
Journals	5″ x 6″

ENGINE TRUCK WHEELS (Front)

Diameter	24¼″
Journals	3½″ x 6″

ENGINE TRUCK WHEELS (Back)

Diameter	24¼″
Journals	3½″ x 6″

WHEEL BASE

Driving	9′ 0″
Total	20′ 8″

WEIGHT

On Drivers	41,995 lbs.
On Truck, Front	9,000 lbs.
" " Back	7,350 lbs.
Total	58,345 lbs.

TANK

Tank Capacity	650 gals.

SERVICE

Passenger and Freight.

To work on curves of 1 ft. in 30 ft.
Radius of curves 300 feet.

BALDWIN LOCOMOTIVE WORKS

Mogul Locomotive

for the

Zululand Railway (Natal)

Class 8-24-D-115

Gauge 3′ 6″

GENERAL DIMENSIONS

CYLINDERS

Diameter	15″
Stroke	18″
Valve	Balanced

BOILER

Diameter	50″
Thickness of Sheets	½″
Working Pressure	160 lbs.
Fuel	Soft Coal

FIREBOX

Material	Steel
Length	70³⁄₁₆″
Width	29″
Depth, Front	53⅛″
" Back	44⅛″
Thickness of Sheets, Sides	⁵⁄₁₆″
" " " Back	⁵⁄₁₆″
" " " Crown	⅜″
" " " Tube	½″

TUBES

Material	Iron
Number	151
Diameter	1¾″
Length	9′ 11″

HEATING SURFACE

Firebox	79 sq. ft.
Tubes	677.28 sq. ft.
Total	756.28 sq. ft.
Grate Area	14.1 sq. ft.

DRIVING WHEELS

Diameter Outside	42″
" of Center	36″
Journals	6″ x 7″

ENGINE TRUCK WHEELS

Diameter	26″
Journals	4″ x 6½″

WHEEL BASE

Driving	10′ 9″
Total Engine	18′.0″
Total Engine and Tender	37′ 9½″

WEIGHT

On Drivers	56,490 lbs.
On Truck	9,200 lbs.
Total Engine	65,690 lbs.
Total Engine and Tender	125,000 lbs.

TENDER

Diameter of Wheels	28″
Journals	4″ x 8″
Tank Capacity	2,400 gals.

SERVICE

Freight.

To work on curves of one foot in thirty feet; radius of curves 300 feet.

BALDWIN LOCOMOTIVE WORKS

Mogul Locomotive

for the

Egyptian State Railways

Class 8-30 D-593

Gauge 4' 8½"

GENERAL DIMENSIONS

CYLINDERS

Diameter	18"
Stroke	22"
Valve	Balanced

BOILER

Diameter	54"
Thickness of Sheets	7/16"
Working Pressure	160 lbs.
Fuel	Coal

FIREBOX

Material	Steel
Length	71¹¹/₁₆"
Width	34⅜"
Depth, Front	72¼"
" Back	70¼"
Thickness of Sheets, Sides	5/16"
" " " Back	5/16"
" " " Crown	⅜"
" " " Tube	½"

TUBES

Material	Iron
Number	192
Diameter	2"
Length	10' 5¾"

HEATING SURFACE

Firebox	112.3 sq. ft.
Tubes	1045 0 sq. ft.
Total	1157.3 sq. ft.
Grate Area	17.0 sq. ft.

DRIVING WHEELS

Diameter Outside	49"
" of Center	43¾"
Journals	7" x 8"

ENGINE TRUCK WHEELS

Diameter	28"
Journals	5" x 8"

WHEEL BASE

Driving	14' 9"
Total	22' 2"

WEIGHT

On Drivers	91,100 lbs.
On Truck	16,450 lbs.
Total	107,550 lbs.

TANK

Tank Capacity	1,440 gals.

SERVICE

Switching.

PHILADELPHIA, GERMANTOWN, AND NORRISTOWN
RAIL-ROAD.
LOCOMOTIVE ENGINE.

NOTICE.—The Locomotive Engine, (built by M. W. Baldwin, of this city,) will depart DAILY, when the weather is fair, with a TRAIN OF PASSENGER CARS, commencing on Monday the 26th inst., at the following hours, viz:—

FROM PHILADELPHIA.	FROM GERMANTOWN.
At 11 o'clock, A. M.	At 12 o'clock, M.
" 1 o'clock, P. M.	" 2 o'clock, P. M.
" 3 o'clock, P. M.	" 4 o'clock, P. M.

The Cars drawn by horses, will also depart as usual, from Philadelphia at 9 o'clock, A. M., and from Germantown at 10 o'clock, A. M., and at the above mentioned hours when the weather is not fair.

The points of starting, are from the depot, at the corner of Green and Ninth street, Philadelphia and from the Main street, near the centre of Germantown. Whole Cars can be taken. Tickets, 25 cents.

LOCOMOTIVE "OLD IRONSIDES" ON THE GERMANTOWN & NORRISTOWN RAILWAY, WITH FAC-SIMILE OF ADVERTISEMENT AND TIME TABLE, WHICH APPEARED IN "POULSON'S AMERICAN DAILY ADVERTISER," NOVEMBER 26, 1832

BALDWIN LOCOMOTIVE WORKS

Locomotives

Of the Nineteenth and Twentieth Centuries

A Paper read by S. M. VAUCLAIN before the meeting of the

New England Railroad Club

February 12, 1901

I T was with great pleasure that I accepted the invitation to read a paper before you to-night. The opportunity of talking to so many representative railway men assembled in meeting is an indulgence not always permitted by my business.

The subject I have chosen will not only be historical, but, in a measure, prophetic: "The Locomotives of the Nineteenth and Twentieth Centuries." I have obtained lantern slides of some of the earlier locomotives from the "World's Railway," by J. G. Pangborn, which are, no doubt, the most authentic illustrations in existence.

Do you realize that we have now about completed the first century of locomotive building? What terrific strides have been made, what adoration and respect have we, the active young men, stewards of the art, for the grand

TREVITHICK'S MODEL, 1800

BALDWIN LOCOMOTIVE WORKS

TREVITHICK'S LOCOMOTIVE, 1803

men of one hundred years ago, fathers of the art to which we are now applying our best efforts! It is one thing to practice, another to create, and as the art of locomotive building is now thoroughly established, the new century, will, I assure you, record such an advance as will astonish the world. As my paper will ignore all that is foreign to the practice of our own country, I must digress a trifle, in order to introduce the locomotive as it appeared in 1800.

Trevithick's model, which is shown on page 119, was made in 1800. Its first journey was on the kitchen table. Toy as it was, it marks the birth of the successful locomotive. Prior to this we had the creations of Newton, Cugnot, Murdock, Nathan Reed, of Salem, Mass. (the first in America), and others, but their machines were all of locomobile species.

Trevithick's model, however, was the germ of his locomotive of 1803, here shown, the first locomotive in the world made to run on iron rails. It is to Trevithick also that we owe the application of high pressure steam to the locomotive.

In 1804 an American, Oliver Evans, a citizen of Philadelphia, a city so proverbially slow that not even the rivers run through it, brings forth his locomotive, really a locomobile, shown on page 121. Could he but visit "Sleepy Hollow" to-day and appropriate the industry spread over the ground once occupied by his humble workshops, the Baldwin Locomotive Works would have a new proprietor. The usual disappointments of the inventor and experimenter were his, however, and even though this machine propelled itself from his workshop to the river, his necessities compelled him to return to his flour mill machinery.

BALDWIN LOCOMOTIVE WORKS

An example is shown, on page 122, of an attempt to increase the tractive power of the locomotive—the invention of Brunton in 1812. We thus record the first effort to improve, by complication, the hauling power of the locomotive—an *ignis fatuus*, which has proven the financial ruin of many well-known inventors of the nineteenth century.

The next most notable advance, shown by the illustration on page 123, was made by Seguin, in 1827, and represents the first application of the multitubular boiler, the standard locomotive boiler of the present era.

Stephenson's "Rocket," however, shown on page 124, was the "real thing," so far as modern locomotive boiler construction is concerned. It was a winner from the start, and assured future success if the ideas of its inventor were persevered in by those who were to follow.

We now turn to America alone to describe briefly the improvement in the art. The "Howard," in 1830, was the first locomotive to be built in America, designed by a civil engineer in the employ of the Baltimore and Ohio Railroad, a corporation then in its infancy.

In the same year Peter Cooper also took a hand in the development of the locomotive, and successfully operated his experimental model, shown on page 125. He was followed by Johnson, Milholland, Winans and others, all destined to shine brightly in the constellation of the nineteenth century inventors.

But it was a jeweler whose great sagacity and business energy enabled him to establish at this early

OLIVER EVANS' MODEL, 1804

BALDWIN LOCOMOTIVE WORKS

date a Locomotive Works to produce, under contract, locomotives for a fixed price. His work was so well done, his many inventions so necessary and useful, and his business policy so accurate and well defined, that his successors have been able, by adding thereto their meagre contributions to the art, to perpetuate his workshop, until to-day in placid Philadelphia we have the largest locomotive works in the world. Its product is familiar to every country in the world, with a capacity of over 1,200 locomotives annually, which is 40 per cent. of the American production, and all an evidence of the sound engineering and business sagacity of its founder.

The "Ironsides," shown on page 126, was a single-driver locomotive, a type which prevails even to-day, and gives most excellent service where the traffic is light.

BRUNTON'S MODEL, 1812

In 1832 the Stephenson link was first successfully applied by James, of New York, to a locomotive shown on page 127.

Horatio Allen's eight-wheel engine, shown on page 128, built in 1832, is the forerunner of the modern Fairlie class of locomotives. The same aversion, however, that to complication—common to all Americans—sealed its fate, but the device marks another step in the art.

The year 1832 records the first application of the Bogie, or four-wheel truck, shown on page 129, which has been generally used in America since that time. Although for many years its use was not favorably considered, in English practice it has been universally adopted in all recent English designs.

122

BALDWIN LOCOMOTIVE WORKS

No reference to the progress of the locomotive would be complete without the "Crabs," of 1832–34, operated so long by the Baltimore and Ohio R. R., and only recently cut up for scrap, after having been in service two-thirds of a century. An illustration of one of these locomotives is shown on page 130.

In 1836 H. R. Campbell designed, and had Mr. Baldwin build, the ordinary eight-wheel engine, since known as the American type, a type that has been able to do more in the development of this country than any other type of locomotive, a type which is perpetuated even to-day, magnificent specimens appearing in every railway station. This was the first distinct advance.

SEGUIN'S LOCOMOTIVE, 1827

The Campbell engine, shown on page 131, immediately suggested the equalizing beams, which were employed on the engine "Hercules," shown on page 132, built in 1837 by Eastwick & Harrison, a Philadelphia firm of locomotive builders. Bolted stubs were also employed on this engine instead of the gib and key. 1837 records the first product of the once famous Rogers Locomotive Works, the recent closing of which is a sincere regret to every true American. This engine, shown on page 133, is remarkable for being the first locomotive sent west of the Ohio River; also the first locomotive to have four fixed eccentrics, and counterbalance weights in driving wheels.

In 1838 we get down to business. The illustration on page 134 shows Norris' supreme effort for the Baltimore and Ohio R. R., and the beginning of a long line of successful engines by various builders.

BALDWIN LOCOMOTIVE WORKS

STEPHENSON'S LOCOMOTIVE, "ROCKET," 1829

It was in 1840 that the first American locomotives were exported to England. These were built by the Norris Locomotive Works, of Philadelphia, now absorbed by the Baldwin Locomotive Works.

In 1842 Mr. Baldwin, harassed by the demands of the railroads for heavier and more powerful locomotives, and further stimulated by the success of his many competitors, patented his flexible beam truck, which enabled him to compete successfully with all builders and secure to himself a monopoly of heavy freight locomotives. An illustration of the Baldwin flexible beam truck is shown on page 135.

In 1844 the first eight-coupled freight locomotive appeared, built by Ross Winans, whose name stands high among those who have advanced locomotive building during the nineteenth century. It was the forerunner of American locomotives for freight service, and is shown on page 136.

In 1846 Mr. Baldwin introduced his eight-wheel connected locomotive, with flexible beam truck, for freight service, shown on page 137. This type was so successful that it was not until 1855 that Mr. Baldwin was finally prevailed upon to abandon this design and place his cylinders horizontally.

In a design for a locomotive with horizontal cylinders Winans once more succeeded where many had failed. In 1846 the engine shown on page 138 was built; it was the first to burn anthracite coal successfully, and thus enabled the Reading Railroad to haul coal to Philadelphia by burning coal instead of wood.

In the same year Septimus Norris and John Brandt brought out the ten-wheel engine, illustrated on page 138. This type has been tenaciously adhered to for freight service during the last half of the nineteenth century, but is now being superseded by larger and heavier power for heavy freights at slow speed.

BALDWIN LOCOMOTIVE WORKS

And again in 1848 Winans astonishes the locomotive fraternity with his "Camel," shown on page 139. Without this design much of the progressive heavy grade engineering of the pioneers would have met with poor success.

The illustration on page 139 shows the more perfect "Camel" of Winans—the "Camel" of 1851—the locomotive of my youth, and for which I still have the utmost respect. One of these locomotives was used by the Baltimore and Ohio R. R. to work a temporary line with a grade of five hundred and thirty feet to the mile, while awaiting the construction of a tunnel, the weight of engine and train aggregating fifty tons. Of course it would not do for the modern mechanical engineer to scrutinize too closely the details of these locomotives, for if this was done the advocate of butt joints and double cover-strips with sextuple riveting would be alarmed at the result of his investigation. The first half of the nineteenth century has thus passed in the development of principles and types, which the last half of the century has been perfecting, introducing and applying a science of manufacturing to keep pace with the advance of the art, recording and sifting the data of practice for the perpetual benefit of all.

In 1851 Wilson Eddy, with whom all New Englanders are familiar, introduced the "Gilmore," shown on page 140, with wheels six feet nine inches in diameter. The many peculiarities of the various productions of Mr. Eddy have caused him to be generally admired and respected. The

PETER COOPER'S MODEL, 1830

BALDWIN LOCOMOTIVE WORKS

placing the cylinders of an eight-wheel engine horizontally, as shown in this design, was thus early established as an assured feature in this type.

It would be slighting one of the most original and progressive men of the middle century did we not show James Milholland's "Illinois," shown on page 141, used on the Reading Railroad in 1852, the first passenger engine to burn anthracite coal. This locomotive was equipped with wrought-iron driving wheels seven feet in diameter, and attained, at times, a speed of seventy-five miles per hour. The cylinders were seventeen inches diameter by thirty inches stroke. These locomotives were the most advanced and mechanically perfect machines produced at that period. They excited much admiration and were largely imitated.

In 1853 Mr. Baldwin abandoned the half stroke cut-off he had used since 1845, and introduced his variable cut-off as shown on page 142. His opposition to the Stephenson link was enduring, and not until the pressure of his associates was brought to bear on him did he consent to its adoption. The plan of casting the cylinders with a half saddle was first introduced by Mr. Baldwin, and naturally so, because his inclined cylinders were bolted to a circular smoke box, and as the cylinders were lowered to the horizontal, the saddle assumed its present shape.

In 1853 Norris built the Phleger boiler, shown on page 143, which was one of the early improvements in the regular locomotive boiler. The Boardman boiler had, however, been made somewhat earlier, but both were unsafe and gave poor satisfaction.

BALDWIN'S LOCOMOTIVE "OLD IRONSIDES," 1832

126

BALDWIN LOCOMOTIVE WORKS

The first attempt to introduce water tube boilers, reversing the order of things and passing the water through the tubes, was made by Dimpfel about the same period. Many of these were built by Mr. Baldwin, but ultimately abandoned on account of the amount of repairs necessary, many tubes having to be removed to get at any one which might be defective.

Although many attempts had been made to provide a locomotive boiler of the wagon-top type, the most notable example was in the "Madison," shown on page 143, built by Rogers, thus providing a model for us all.

In 1860 the war clouds were rising, and during the next five years much that could have been done was left undone, to attend to more important matters. One thing, however, was

LOCOMOTIVE WITH STEPHENSON'S LINK, 1832

established, and that was the necessity for the rapid construction of locomotives, to facilitate the development of our country; consequently renewed energy was thrown into the work, and the superiority of the American locomotive was assured. Soon our creations were beating time in all the countries of the world, and to-day Americans are feared by all foreign builders of locomotives, even in their home markets.

In 1861 the "Mogul" locomotive appeared, built by the Baldwin Locomotive Works for the Louisville and Nashville R. R. This locomotive, shown on page 144, is undoubtedly the first Mogul with a flexible truck. It is a ten-wheel design, with a two-wheel sliding Bissell truck in front of the cylinders.

Moguls seemed to supply a long-felt want. Mr. John P. Laird, at Altoona, on the Pennsylvania R. R., converted a number of the Camel engines on that line into Moguls, even going so far on the old "Seneca,"

BALDWIN LOCOMOTIVE WORKS

HORATIO ALLEN'S DOUBLE-ENDER LOCOMOTIVE, 1832

shown on page 144, as to omit the second pair of drivers and add the truck. If he had only allowed those wheels to remain he would have had the honor of being the father of the consolidation type, now such a favorite, and, indeed, I feel that he is entitled to it anyhow, because two engines, as shown on page 145, were built by him at Altoona during the year 1865, and were cut up in 1869. You will notice that the vexed question as to the use of flanged tires on all drivers was a settled fact, even at that early day. You will also recognize in this engine the Consolidation engine of to-day. These engines were used for mountain pushers.

In 1861 steel fireboxes of English make were introduced, which failed, but in 1862 steel fireboxes of American manufacture were placed in engines Nos. 231 and 232, on the Pennsylvania R. R., duplicate of that shown on page 145, and since that time steel fireboxes have been the American standard. Steel driving wheel tires were also introduced at this time.

In 1866 the "Consolidation," shown on page 146, was built by the Baldwin Locomotive Works from designs and specifications of Alexander Mitchell, for the Lehigh Valley Railroad Company, and thus was firmly launched the first of a long line of successful locomotives that have penetrated almost every country on the globe where heavy traffic is found. Mr. Mitchell has been appropriately referred to as "the father of the Consolidation locomotive." During 1870 to 1879 the principal efforts of our engineers were directed to the development of narrow gauge power, of which I will not speak. In 1870, however, the practice of shrinking steel tires on driving wheels without shoulder or set-screws was inaugurated and is now general.

BALDWIN LOCOMOTIVE WORKS

Additional types of locomotives, such as double-enders, shown on page 147, appeared about this time; also, in 1874, the first compressed air locomotive, for street-car work, was made by the Baldwin Locomotive Works.

In 1876 the exhibition at Philadelphia gave fresh impetus for locomotive improvement, and many enlargements of previous practice were introduced with great effect on various lines.

In 1877 Mr. John E. Wootten introduced the wide firebox projecting over and above the driving wheels, as shown on page 148. To-day this form of firebox is being used to a very great extent. Also many modifications of this design are being extensively introduced. One of these modifications is shown on page 149. The inventor passed beyond before the world was ready to acknowledge the great service he had performed.

In 1880 high speed commenced to worry the locomotive brethren, and the engine shown on page 150, illustrates what was considered necessary, and was the forerunner of our present most successful engines for this service.

Although many locomotives with more than four pairs of driving wheels coupled had been built, 1885 marks the introduction of the "Decapod" type, or locomotives with five pairs of coupled wheels and a pony truck, as shown on page 150. Engines of this type have frequently been built, but owing to the constantly increasing weight of rail used, the Consolidation engine has so far been able to supply the demands of service.

FIRST LOCOMOTIVE WITH FOUR-WHEELED TRUCK, 1832

BALDWIN LOCOMOTIVE WORKS

The extended wagon-top boiler now so generally used was introduced in 1886. It was designed by Mr. William L. Austin, of the Baldwin Locomotive Works, for Mr. Sample, of the Denver and Rio Grande R. R., shown on page 151. It was only new, however, in that it was a radial stay extended wagon-top boiler, extended wagon-tops having been used by Baldwin thirty years prior, when combustion chambers with crown bars were in vogue.

In 1887 steam, as an agent of locomotion, encountered electricity, and many splendid specimens of electric locomotives have since been built; those operating in the Baltimore Tunnel, of the Baltimore and Ohio R. R., shown on page 153, being the most powerful. The electric locomotive, yet in its infancy, has accomplished much, but has not earned consideration as a substitute for steam locomotives, except where the service is of a special nature, such as tunnels, mines, factories, and on elevated railroads. An electric locomotive for mine haulage is shown on page 152.

ROSS WINANS' "CRAB," 1832-34

In October, 1889, a successful compound locomotive, shown on page 153, was introduced by the Baldwin Locomotive Works, and three followed in 1890. Since which time nearly 2,000 have been built, and are now sought by engineers in all parts of the world.

Successful two-cylinder compounds appeared about the same time, many of which are now in successful operation. One of these, in use on the Old Colony R. R., is shown on page 154,

The "Columbia" type, shown on page 154, next appeared, in 1892, designed by Mr. William P. Henszey,

130

H. R. CAMPBELL'S EIGHT-WHEEL LOCOMOTIVE, 1836

who for nearly fifty years has been connected with the engineering department of the Baldwin Locomotive Works, and who has designed more locomotives than any other engineer living, and is now one of the senior members of that firm. This new idea in locomotives was exhibited at Chicago, in 1893, as the ideal of a high speed locomotive. It was tested thoroughly and demonstrated the wisdom of the designer, but gave place, in 1895, to his more favored design, the "Atlantic" type. The first "Atlantic" type locomotive, shown on page 155, was constructed by the Baldwin Locomotive Works. This has proven the most successful and advanced high-speed locomotive so far produced, and its popularity has been well earned, having experienced great opposition from various locomotive builders and railway men. To-day it is used by all the leading lines in the country, and is being built by other builders under such names as the "Northwest" and "Chautauqua" types.

These engines of the "Atlantic" type have the highest speed records of the world, and are now being copied by many of the leading lines on the continent, as well as by some English lines.

Just at the close of the century we have successfully introduced what is known as the "Vanderbilt" boiler, shown on page 156. The abolition of stays, etc., is a great advance, but three or four years must elapse before the conservatism of our motive power departments will permit its acceptance. This boiler was designed and introduced by Cornelius Vanderbilt, and gives great promise for the future. The rapidity with which the firebox can be renewed, and the absence of usual repairs on a locomotive, commend it. The fact that the

BALDWIN LOCOMOTIVE WORKS

EASTWICK & HARRISON'S LOCOMOTIVE, 1837

boiler has been promoted by Mr. Vanderbilt is, perhaps, somewhat of a handicap; as were it a Brown boiler or a Smith boiler it might receive more prompt recognition. For full information, see paper read by Cornelius Vanderbilt before the American Society of Mechanical Engineers, January, 1901.

Before closing the record of a century of labor and advancement, such as the world never had before, a few samples of modern locomotives may be referred to, of which a full description can be had hereafter by those interested. See pages 157 to 163.

To the twentieth century, just born—and the best one of them all—we look for rapid strides in the further development of the locomotive. The construction of large capacity cars, makes it possible to greatly increase the train tonnage without increasing the length of the train. They also increase the terminal facilities of the large roads to such an extent that it is difficult to foresee how far this increase in car capacity will be carried. One thing is certain, however, that many of the locomotives at present employed, are inadequate and unprofitable to operate. Already their size is being increased to the maximum weight that rail and bridges will permit. The increase will, however, be in periods, probably five years in extent, or so long as paying freight can be hauled. When this ceases to be the case, then increased strength of rail and bridges will be provided to carry the heavier locomotives.

There will be a battle royal, sooner than many expect. The improvement of the locomotive will embrace the further development of those features invented in the previous century, compounding of all locomotives upon some system now used, or yet to be invented, will be almost universal, the wide fire-

BALDWIN LOCOMOTIVE WORKS

box and tubular boiler will be carried to the limit of human ability to manage it. This will give place to the water tube boiler, especially for high speeds. Who is destined to be the instrument of its introduction? Already bright minds are employed in designing a boiler of this description, which can be placed on our arrangement of cylinders, underframing, wheels and machinery—a system that will give three times the heating surface for an equivalent weight. Higher pressures will then be common, and we all may live to see triple and even quadruple expansion locomotives almost noiselessly performing their work. High speeds, consistent with safety, will be used for all trains carrying human freight, but long and heavy express trains will be handled with facility by the improved high pressure compound locomotives of that period. The loading gauge of our trunk lines will not prevent doubling, or even trebling the power of locomotives for freight traffic. A Double Bogie engine, similar to those used abroad, but on the American idea, will be employed, thus reducing it to an almost perpetual operating machine, any part of which can be removed in a short time and a duplicate substituted.

The lives of great men are being spent in an endeavor to solve the problem of electricity as a motive power. Water power, where available, is now being employed for supplying current; electricity, as a motive power, will steadily gain friends; large plants will be erected for this purpose. One of the greatest features of this problem, if not the greatest of all, relies upon the use of gas, made at the mines and sent through pipes to the power plants placed at intervals along the line, and used direct in gas engines of huge units for gener-

133

ROGERS LOCOMOTIVE, 1837

BALDWIN LOCOMOTIVE WORKS

ating the current necessary to operate the traffic of the road. Can it be that the system of multimotors now used by various elevated railways, will be employed with success on the rolling stock of our trunk lines? Perhaps so, but it is my opinion that for our trunk line traffic, electricity will not be used until it can be generated economically on the locomotive itself. It is in this direction we look for success, as we now use up but a small per cent. of the calorific power of coal in our best steam locomotives. If, by the invention of a method whereby almost the entire calorific power of coal is converted into electric current directly, and not through our present system of engines and generators, we will be compelled to abandon the steam locomotives, then the electric era will really be at hand. As the electric locomotive occupies relatively the position that steam locomotives did a century ago, those who live in 2001 may merely have recollections of the wonderful events of our present age.

Have any of you as yet considered to what extent the pneumatic tube will be employed to expedite transportation now entirely dependent on locomotives? Has anybody watched the long lines of coal cars on their way from the mines to the coast, and the same cars returning empty? If the weight of a car is 25 per cent. of the gross load, we have more than 50 per cent. loss, or non-paying freight, when we consider that the empty train requires quite as much power to haul it up into the interior as was expended taking it to the coast. Is it not possible? Will it not be accomplished? And just as the miles of cars loaded with oil, seen in former years, have disappeared, and that commodity sent hundreds of

NORRIS' LOCOMOTIVE, 1838

BALDWIN LOCOMOTIVE WORKS

miles through a tube in the ground, will coal, grain and ore be sent speeding through tubes to central depots for local distribution.

Had any man been so bold as to prophesy in 1801 our present status in relation to the arts and sciences, he would have been looked upon with suspicion, if not actually confined. Therefore I object to describing in detail the various stages in the future development of the locomotive, and will allow my friends the golden opportunities at hand for material prosperity, fame and a place in history.

BALDWIN FLEXIBLE BEAM TRUCK

BALDWIN LOCOMOTIVE WORKS

MR. W. P. APPLEYARD. Mr. Chairman, I move a most cordial vote of thanks to Mr. Vauclain for his very interesting paper.

THE CHAIRMAN. Mr. Vauclain desires me to say that he will answer any questions that any one may wish to ask. I now give you the opportunity. There will be no topical discussion to-night.

DR. J. B. THORNTON. Mr. Chairman, I would like to ask the speaker of the evening to tell us how some of our American locomotives compare to-day with those which are in use on foreign railroads. We know we are sending a great many of our engines abroad, and we would like to have a few words upon that point.

MR. VAUCLAIN. The principal reason why American locomotives are so persistently sought by foreigners is the low price at which they are sold, and their reliability and flexibility. Foreign-built locomotives are said to be too rigid for the colonial work. Our locomotives are sent out into the colonies, and they are operated on railroads that somewhat closely con-

ROSS WINANS' EIGHT-COUPLED LOCOMOTIVE, 1844

form to the railroads in the United States, and therefore in order to derive the best results, it is necessary to have as flexible machines as possible, and which heretofore our competitors abroad have been loath to build. The progress that Americans have made in the introduction of their locomotives into foreign countries has led many of the foreign workshops to copy, so far as possible, the flexible features of these particular

BALDWIN LOCOMOTIVE WORKS

locomotives, and in the several markets where our locomotives are sold, more especially in South America, we encounter what is called an improved American locomotive, which is simply the English locomotive provided with equalizing beams that are flexible.

MR. DEAN. Mr. Vauclain, I have been told recently that a firm in Berlin was equipping itself to build American locomotives, that they were importing American machinery, with an American superintendent of the work. Do you know anything about that?

MR. VAUCLAIN. That has been rumored, but I think that rumor was caused by the introduction of some American locomotives of Baldwin build, some compound locomotives into Bavaria, two freight locomotives, as samples, and these locomotives are being carefully tested, and are to be modelled after for future locomotives for their freight traffic. These locomotives did so well that the management decided to order two passenger locomotives, thinking that by so doing they could gain enough information, enough

BALDWIN EIGHT-COUPLED LOCOMOTIVE WITH FLEXIBLE TRUCK

American ideas to enable them to improve their passenger rolling stock. These locomotives have also been shipped, and are at this time being erected. We were practically notified at the time that we need not expect further orders, that these locomotives were simply ordered as samples, and that all the new features in them would be copied and possibly adopted in their future engines.

BALDWIN LOCOMOTIVE WORKS

WINANS' "DELAWARE," WITH HORIZONTAL CYLINDERS, 1846

MR. F. W. DEAN. I think, Mr. Chairman, that one of the most novel ideas that Mr. Vauclain has mentioned this evening, is that of transmitting solid articles through pipes, such as coal and grain. The pumping of petroleum has resulted in great economy, but to me it is entirely novel to think of sending grain or coal in a similar way. I have just been told by Professor Allen that the scheme has been suggested of pumping a mixture of coal and water, making the water carry the coal. I do not imagine that it would be possible to carry solid matter like coal unless a liquid medium like water should be used to do it; that is to say, I could hardly imagine that coal could be blown with air, although grain might possibly be. In woollen mills it is a very common thing to blow wool from one building to another through pipes by means of fans.

MR. VAUCLAIN. I might say for the benefit of Mr. Dean that experiments have already been made in this line. It has been demonstrated that coal can be successfully put through the tube, but the unfortunate pulverizing of coal would require buttresses at the other end for properly checking it, and that was one of the great difficulties that the inventor met with. But in the early days of locomotive building obstacles almost as great were

NORRIS' TEN-WHEEL LOCOMOTIVE, 1846

138

encountered, and I think it is perfectly proper for us to expect that some twentieth century inventor may, by mixing fuel with some other agent, successfully carry it through tubes.

THE CHAIRMAN. Has any one else a question that he would like to ask Mr. Vauclain?

MR. SMITH. Mr. Chairman, I would like to ask Mr. Vauclain about the introduction of engines from the Baldwin Locomotive Works on English railroads. I believe the Midland Railroad uses American locomotives.

MR. VAUCLAIN. We built last year in all seventy locomotives for three leading lines in England. We built

ROSS WINANS' ORIGINAL "CAMEL," 1848

thirty for the Midland, twenty for the Great Northern, and twenty for the Great Central. These locomotives are giving very good satisfaction indeed. When they were first put in operation the usual exception was taken to them, that they were not economical as compared with the English locomotive; but on the other hand, while they may not be so economical in fuel, they were used on trains considerably below their capacity. It was found that they would take additional train loads, and make any speed that was required. The question of fuel

ROSS WINANS' "CAMEL," 1851

economy between the two types, the English and American type, is a very simple matter. If we should reduce the throttle of our engine and check the flow in their steam passages and steam pipes, we could so reduce the fuel consumption per engine mile as would compare well and favorably with some locomotives built in England and in regular service there. That is not our intention, however, and each day that these locomotives run in England they are being better thought of by the English people, which we think is quite a concession.

MR. A. B. AVERILL. Mr. Chairman, I would like to ask the speaker of the evening to what extent they have built locomotives for burning petroleum, and also what success those locomotives meet with?

MR. VAUCLAIN. We have built in all about four hundred and fifty locomotives for burning petroleum, and we have been entirely successful with every one of them. The device that we use is one that has been in use for many years in Peru on the Oroya road. It consists of one pan immediately over another pan. The upper is the pan over which the oil flows.

WILSON EDDY'S LOCOMOTIVE, "GILMORE," 1851

The under pan has a throttle at its mouth, and is the steam pan. The heat from the steam liquefies the oil, and as the oil drops over the edge of the pan or burner, it is caught by the steam jet, which is probably three or four inches in width, and about the thickness of a piece of paper, and it is spread out over the interior of the firebox. No further preparation is necessary on a locomotive, other than to remove a portion of the gratebars. There is a little firebrick near the front end to protect the throat sheet. The burner is fastened to the bottom of the mud ring, and at such an angle as will allow it to shoot up into the firebox.

BALDWIN LOCOMOTIVE WORKS

Most of our locomotives that we have built for the Russian government in the last three or four years have been able to burn refuse petroleum, and are all giving most excellent satisfaction.

MR. F. W. DEAN. Mr. Chairman, so far our discussion of Mr. Vauclain's paper has related to novelties which are more or less startling, but we are not discussing what Mr. Vauclain's paper was more particularly intended to bring to our attention, that is, the tendency of locomotive practice. I think that the most important feature of the tendency of present practice, is the recognition of the Atlantic type of loco-motive. The fact that other builders than the Baldwin Locomotive Works have begun to build them shows such recognition. It is my belief that this will be the most conspicuous type of passenger locomotive to be seen on roads where loads of any great weight are to be hauled at high speed. Mr. Vauclain passed over that matter some-what hastily, and perhaps did not explain its advantages, but, as I understand it, the advantages of the Atlantic type of locomotive are that the driving wheels are entirely in front of the firebox, and that the firebox can thereby

MILHOLLAND'S PASSENGER LOCOMOTIVE, 1852

be made wide. More grate area can be obtained than in the ordinary locomotive without excessive length of firebox. Great length of firebox, means difficulty in firing. The Atlantic type not only gives great width of firebox, but also considerable depth, because there is underneath a small pair of trailing wheels. The trailing wheels serve also to carry the extra weight of the large firebox and boiler. In this way boiler capacity is very much increased. The tendency in the large locomotives in the last few years has been to

BALDWIN LOCOMOTIVE WORKS

increase the heating surface, while the grate area remained constant, but by the use of the Atlantic type, the grate of the locomotive is very much extended, and keeps pace with the heating surface.

Very wide fireboxes, as Mr. Vauclain has shown, were used with eight-wheeled, four-coupled anthracite burning locomotives, long before the Atlantic type of engine was brought out, by extending a shallow firebox over the rear driving wheels. Large wheels, however, render the design impracticable.

The New York Central R. R. now has the heaviest engines of the Atlantic type so far brought out.

There is another thing that Mr. Vauclain has not touched upon, viz., piston valves. I think that most of us have the idea that piston valves are not successful pieces of apparatus, and I should be glad to have his opinion of them. I have had some doubt about them and should be glad to have it removed. I should be glad also to know from Mr. Vauclain whether measurements show that by the use of piston valves the areas of indicator diagrams have been increased in comparison with those from slide valve engines, the locomotives being of the same general character and size.

MODEL OF BALDWIN CUT-OFF

MR. VAUCLAIN. In regard to the piston valve as used in the locomotive, I desire to say that in our four-cylinder compound we find that it is necessary for us to use a piston valve. The piston valve was objected to by many. It was one of the principal things on the engine that it was thought would condemn it, and would prove a source of weakness in its design. The object in using the piston valve was to combine two valves in one, and at the same time provide a receiver for the high pressure cylinder to temporarily exhaust into before the low pressure engine was ready to receive its charge of steam. The use of piston valves

BALDWIN LOCOMOTIVE WORKS

on the simple engine is quite a different matter, however, and I have so far in my experience failed to see the necessity of using piston valves on a simple or single expansion locomotive. The piston valve on a single expansion locomotive, whilst it may have some advantages, has disadvantages. The balance of favor, however, from my observation and from my point of view, still rest, with the slide valve, if properly designed. The piston valve is said to operate more steadily and easily, but I fail to see it, provided the slide valve is properly

NORRIS' LOCOMOTIVE, WITH PHLEGER BOILER, 1855

adjusted. There is a serious source of inconvenience in the use of the piston valve, as it seals the cylinder at a time when the engine may be working water, thus causing damage to the cylinder heads, the water not being able to escape. We have met with this difficulty in compound locomotives where we have been compelled to use the piston valve, and we have provided means for overcoming it, that is to relieve the cylinder of excessive pressure that would be caused by the entrapped water. With a single expansion locomotive in introducing piston valves we have introduced something that will require the introduction of something else to avoid an accident.

ROGERS' LOCOMOTIVE, WITH WAGON-TOP BOILER, 1855

143

BALDWIN LOCOMOTIVE WORKS

BALDWIN MOGUL, 1861

Referring to Mr. Dean's mention of the Atlantic type of engine, I would say that in freight service the measure of the efficiency of the locomotive is its maximum tractive effort, the ability of the cylinders to utilize all the weight on the driving wheels. That is not so much a question of boiler capacity, but in passenger service I may state that it is entirely the opposite. It is there a question of steaming capacity, not maximum tractive effort. It is therefore considered better and wiser engineering for that class of traffic to build a locomotive of the Atlantic type rather than that of the ten-wheel type which has been used by many roads, as in the Atlantic type we are able to get all the weight on the drivers that is required. Naturally the heaviest passenger trains can be hauled at high speeds except where the grades are exceedingly heavy, and at the same time the locomotive is able to use the very largest boiler possible, and by placing the firebox back of the main driving wheel over the trailer, we are at liberty to make the grate anything that we see fit, up to one hundred square feet, obtaining the ratio of grate to heating surface that is most desirable. Thus it is apparent that this grate would allow a locomotive boiler to be constructed with six thousand feet of heating surface, if required.

Mr J. H. Graham. Mr. Chairman, Mr. Dean has raised a very interesting point in relation to the Atlantic

144

LOCOMOTIVE "SENECA" RECONSTRUCTED

BALDWIN LOCOMOTIVE WORKS

type. There is no question but that it grew out of an accident much after the manner many things on railroads have. You all know our present track gauge of four feet eight and a-half inches was used because the wagon gauge or tracks were that measurement. It would have been vastly more mechanical and common sense to have made it four feet six inches or four feet nine inches, or better, five feet, and although during the short history of steam railroads there

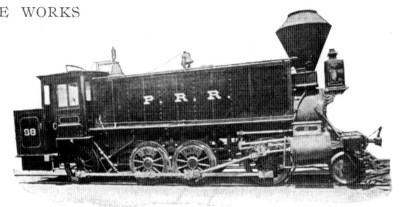

EIGHT-COUPLED LOCOMOTIVE, WITH LEADING TRUCK, 1865

have been tracks with gauges varying from seven feet to thirty inches, yet the earliest lines seem to have held, and we have our present standard.

That the Atlantic type was the result of careful plans no one will doubt, but that it was not an evolution from some other man's effort I cannot allow, for I know that in June, 1880, Wootten placed a pair of forty-two inch wheels under the wide firebox of a single driver engine built for the Philadelphia and Reading. In July, 1884, Alexander Mitchell, M. M. of the Lehigh Valley, built an engine with four pairs of drivers and a leading pair of small wheels and a pair under the very back edge of the firebox; again in February, 1886, Strong shows an engine of the present ten-wheel type with a single pair of small wheels under the firebox. All these were germs.

It certainly would be an accident if some designer making a standard engine with too much weight on the drivers for the bridges should for expediency or safety place a small pair of wheels under the firebox to distribute the weight.

145

BALDWIN LOCOMOTIVE, WITH STEEL FIREBOX, 1861

BALDWIN LOCOMOTIVE WORKS

MR. VAUCLAIN. Mr. Chairman, I would like to reply to that. The speaker has said that in his opinion the Atlantic type of locomotive was an accident. I beg to differ. The time came round when we were given the problem of hauling a certain number of cars at a rapid speed in miles per hour, with a weight on the driving wheels of seventy-two thousand pounds. The tractive effort required, of course, was low, but the steaming capacity required was high. We had never built a Columbia type or American type of locomotive up to that time that we could guarantee to perform this work. The problem was given the utmost consideration, and it was decided to modify the Columbia type of locomotive, adding a four-wheel truck, lengthening the boiler to get the necessary heating surface required, which was twenty-four hundred square feet, and at the same time not exceed the limit of weight on the driving wheels over seventy-two thousand pounds. It was not an accident. It was done with intent. The engines were successful, the guarantee was met, and we were paid.

BALDWIN CONSOLIDATION LOCOMOTIVE, 1866

PROFESSOR ALLEN. Mr. Chairman, as I understand the matter, people wanted a locomotive of practically that type. That form of locomotive was built which rendered that type substantially necessary.

MR. VAUCLAIN. Yes, sir, you are correct. Accidents of that kind are happening daily.

MR. GRAHAM. Mr. Vauclain has just informed me that the Atlantic type was built in 1895. Although he objects to my statement that it was an accident, nevertheless I recall the Miller engine that ran for years on the New York, Providence and Boston R. R., and was built in the latter 80's. L. M. Butler was superin-

BALDWIN LOCOMOTIVE WORKS

tendent of rolling stock. This was a standard engine with a very short wheel base and heavy boiler. When they placed it in service they found the weight on the drivers was too great for the bridges, so they pulled her in and placed a pair of small wheels under the firebox, making the present Atlantic wheel arrangement, although the leading wheels were the drivers.

Many and many a time have I gone over the road when this engine was working the 10 A. M. Shore Line express with young Butler at the throttle, and the way that engine wheeled the heavy trains to New London and return would have pleased the most fastidious. Every time I saw her backed up to the train I would remark what a splendid opportunity for a large grate, and all that happened years before the Atlantic type was produced under that name or ever dreamed of.

Whoever claims the credit for the Atlantic type must admit he had some excellent material to draw upon, and had the Wootten, Mitchell and Strong designs, as well as Gooch's and other British designs as far back as 1845 and 1850.

BALDWIN DOUBLE-ENDER LOCOMOTIVE, 1871

MR. VAUCLAIN. Mr. Chairman, that does not answer the point; the cases are not parallel. The locomotive referred to by the speaker was an accident, but it was not an Atlantic type of engine. The same accident occurred on the Chicago, Milwaukee and St. Paul Road. They had a ten-wheel engine built by another builder, which had a pony wheel, a rigid wheel, however, placed under the back end to relieve the

BALDWIN LOCOMOTIVE WORKS

driving wheels of weight. The introduction of a wheel back of the driving wheel in some shape was an accident; it was a necessity to provide a locomotive of a certain type to meet a certain guarantee by relieving the weight on the driving wheels, so that the builder could get the locomotive off his hands.

Mr. W. F. Ellis. Mr. Chairman, I would like to ask a question, how the Atlantic type of engines compares with other types now in use for fast running around curves, the alignment of such curves to be left true. Before Mr. Vauclain answers this question, I would state when employed on the Providence and Worcester Railroad, after its lease to the Stonington Railroad, I rode at high speed one Sunday around its sharp curves, sitting on the fireman's seat of the engine that has been compared to the Atlantic type. This engine was not adapted to fast speed on roads where there are sharp curves as was shown by this trip.

Mr. Vauclain. The Atlantic type of locomotive for short curves, that is for roads where great curvature is encountered, is just as satisfactory as a ten-wheel locomotive. The loco-motive is constructed entirely as the ten-wheel locomotive is constructed. The controlling wheels are flexible, the driving wheels steadier. Any one running these locomotives over crooked roads will bear witness that they are even more comfortable than a ten-wheel engine. The reason for that is that the trailing wheel carries no connecting rod. There is no driving box or anything of that sort underneath the engineer. He is away back from the main wheel; he is back from the rods; he simply gets the motion

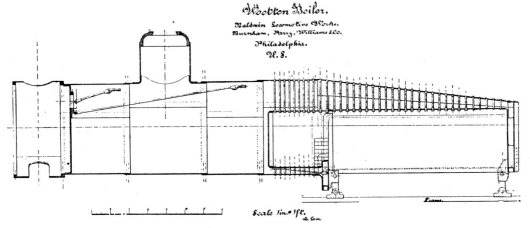

WOOTTEN BOILER, 1877

of the springs. I have ridden on these locomotives at a speed of over ninety miles an hour, have kept it up mile after mile. I have ridden around curves at high speed, and experienced no discomfort. The advantage, as we figure it, is that the centre of gravity is set high, and we have experienced no difficulty whatever.

THE CHAIRMAN. I would like to say a few words in regard to the Atlantic type of engine. Some four or five years ago I went to Philadelphia. I was then quite interested in locomotive work, and I rode several times from Camden to Atlantic City on one of the Atlantic type engines. I remember very well the first trip that I made on the engine. For a number of miles we made a mile in forty-two seconds, and I remember a curve that was ahead of me, and at the rate of speed we were going, I felt somewhat as Mr. Ellis said the engineer and he did on the engine that they rode on, but I was agreeably disappointed. The engine took the curve with just as much ease as it did the straight line. I never rode on a better running engine than on that same Atlantic type engine. I rode on a ten-wheel engine on the Baltimore and Ohio, making about the same time, and the Atlantic type was the more comfortable of the two. The Atlantic type of engine was a four-cylinder compound, and I would like to ask Mr. Vauclain if he would tell us what his experience has been as to the

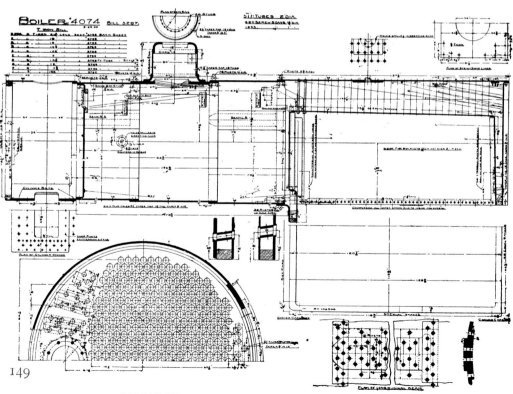

MODIFIED TYPE OF WOOTTEN BOILER, 1900

149

BALDWIN HIGH-SPEED LOCOMOTIVE, 1880

expense of repairs on four-cylinder compounds as compared with a single engine. I was told by the engineer of the engine that I went down to Atlantic City on, that that engine had been on the road for eight months, and had not been in the shop, and had made the trip daily to Atlantic City and back.

MR. VAUCLAIN. Mr. Chairman and gentlemen, in regard to repairing the four-cylinder compound, it is apparent to you all that the man who purchases the locomotive has bother with it more than the man who built the locomotive, but nevertheless, the man who built the locomotive is sure to hear about it if anything out of the ordinary happens. When you consider the fact that last year we built nearly five hundred four-cylinder compound locomotives, and sent them all over the world, principally to American railroads however, and that the originator is still alive, there certainly could not have been very much more than ordinary repairs to those locomotives. On the Chicago, Milwaukee and St. Paul Railroad, a road that has used our compound locomotive for several years, all the locomotives are built on this principle. There they find an actual economy in repairs on four-cylinder compound locomotives over and above what they experienced with their single expansion locomotives. This was not arrived at by comparing two locomotives, but has been arrived at by the repair accounts from year to year. The first test of repairs, however, was between one of the very first locomotives that we had built as against nine single expansion engines, built at the same

150

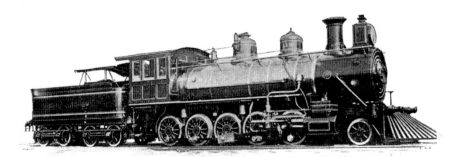

BALDWIN DECAPOD LOCOMOTIVE, 1885

BALDWIN LOCOMOTIVE WORKS

time, exactly similar in all other respects. By the time that the nine simple engines had passed through the repair shops, the superintendent of motive power ordered the compound in to be repaired, and to do such repairs to her as would put her in first-class condition, and turn her out on the road again for service. After that was done the entire repair account of these locomotives from the time they came on the road until they had passed through the shop was tabulated, and to the surprise of the superintendent of motive power, the one that cost least was the compound engine. From my point of view I would expect a compound locomotive to cost a little more for repairs than a single expansion locomotive. I base this upon the presumption that compound locomotives carry higher pressure of steam than the ordinary simple engines do, and they have two more cylinders than the ordinary single expansion engine has, which means more joints, more packing and more piston rods. On the single expansion engine, the cylinder repairs are a very small proportion of the total repairs necessary, and as compared with the compound engine, all other things being equal, I should place a very small percentage of total repairs of the loco-

LOCOMOTIVE, WITH EXTENDED WAGON-TOP BOILER, 1886

motive against the compound cylinders or those parts affected by compounding. That, of course, gives no credit for the reduced repairs with the boiler or the firebox, with the less demand made upon it, nor does it give any credit whatever for the reduced wearing of the valve motion and other parts which are directly benefited by the mechanism employed to operate the valves. I think that the members can safely feel that the repairs to a compound locomotive are not sufficiently greater than those of a single expansion engine to warrant the railroads with which they are connected to continue using single expansion locomotives. The economies obtained in other directions are so large that, from my point of view, no railroad can

BALDWIN LOCOMOTIVE WORKS

economize to a greater extent in its motive power department, than by substituting some class of the compound type locomotive for its motive power.

MR. DEAN. Mr. Chairman, there is another point of locomotive practice that might become a feature of twentieth century practice, and that is the Vanderbilt firebox, about which Mr. Vauclain has spoken. That firebox, as you well know, is a corrugated cylinder. It is self-supporting for external pressure, and requires no stay bolts. The saving of stay bolts is very important, both in first cost and in repairs. It is said to save quite a large percentage of fuel in comparison with a locomotive that has the ordinary firebox. That is something that I cannot quite understand. While I am willing to admit the virtue of the firebox, I cannot see why it should save fuel. I hope that Mr. Vauclain can tell us the reason for it.

MR. VAUCLAIN. We have constructed several Vanderbilt boilers and placed them on locomotives, and they have met with success. It has been demonstrated that this firebox possesses an advantage.

BALDWIN ELECTRIC LOCOMOTIVE FOR MINE HAULAGE

Recently, Mr. Vanderbilt desired to read a paper before the American Society of Mechanical Engineers, and had no recent data for his paper. I told him that I had intended to make some tests of the Vanderbilt engine on the New York Central, if I was permitted to do so, and he said that he would be very glad to arrange for it; therefore I wrote Mr. Waite, and he very kindly gave to my representative permission to make such tests as he desired. The test made on the Vanderbilt engine, in comparison with a sister engine, exactly similar in every respect, with the exception of the size and shape of the firebox, showed a

BALDWIN LOCOMOTIVE WORKS

ELECTRIC LOCOMOTIVE, BALTIMORE TUNNEL

marked economy in fuel, reaching practically ten per cent. This caused surprise; it was a new turn of affairs. It was, as Mr. Dean said, unlooked for and unexpected. Internal examination of one of these boilers developed the fact that no scale of any account had accumulated on the crown sheet or on the side sheets. Of course it is plain that where the crown leaves off the side sheets begin, but there was no scale at all on the upper surface of the firebox, whereas on the crown sheets of the other locomotives a very thick scale had formed; sediment had settled and had not been washed off. The firebox of the ordinary engine was but forty-two inches wide, and was full length, I think ten feet in round figures, whereas in the firebox of the Vanderbilt boiler the grate was about fifty-four inches wide, having a wider grate, and we, therefore, did not need so long a grate. The fire doors were rather high, so that a man had no difficulty in looking at his fire and seeing that the fuel was properly distributed. A portion of the economy, therefore, is attributed to the absence of scale in the firebox and the fact that the grate surface was more nearly square and enabled the man to distribute his fuel to better advantage, and also prevented the fireman from carrying too heavy a fire; whereas in the other engines he can carry the fire as he pleases, and we all know that where it is possible to carry a light fire and have a free circulation of air through the grates, getting the proper amount admitted, and admitted at the right time and in the right place, the combustion is far superior

153

VAUCLAIN FOUR-CYLINDER COMPOUND LOCOMOTIVE, 1889

to what it is where we have a firebox that is choked with fuel too deep and too thick and improperly distributed.

MR. JAMES P. MANNING. Mr. President, we were speaking a few minutes ago about comparisons of simple and compound engines, and the subject recalls a test that was made a few years since on the New York and New England Railroad. That company purchased some Vauclain

TWO-CYLINDER COMPOUND LOCOMOTIVE, OLD COLONY R. R.

compounds. After they had been in use for some months the general superintendent reported to the president that they were a poor investment, that they burned more coal, made less miles, poorer time, and were more expensive to keep in repair than engines of the ordinary type. The report was duly forwarded to Philadelphia, and representatives of the Baldwin concern (Mr. Vauclain was with them) came on with fire in their eyes and demanded an investigation. It was held, and from the road's performance, train sheets and shop records, it was found that the new engines burned less coal per car mile, hauled more cars, made better time and cost less for repairs than the simple ones, with which they were compared. It must, however, be admitted that the repairs of the compounds, although comparatively inexpensive, were of a particularly

vexatious nature, likely to keep the engine-house foremen in a state of mind not at all favorable to the new type.

MR. E. G. DESOE. Mr. Chairman, I would like to ask Mr. Vauclain what per cent. of passenger engines built are compounds, also what per cent. of freight are compounds?

BALDWIN "COLUMBIA" TYPE LOCOMOTIVE, 1892

BALDWIN LOCOMOTIVE WORKS

BALDWIN "ATLANTIC" TYPE LOCOMOTIVE, 1895

MR. VAUCLAIN. Replying to that question, I, beg to say that the percentage of compounds for passenger service was about the same as the percentage of compounds that were built for freight service. We build more freight locomotives than we do passenger locomotives, but the percentage of freight locomotives, equipped with compound cylinders, is practically the same as the percentage of passenger locomotives, built and equipped with the compound cylinder. There are some roads which have their passenger locomotives equipped with compound cylinders, but not their freight locomotives. On the other hand, there are railroads which have all their freight locomotives built compound, but none of their passenger locomotives. It is only a question of time, from my point of view, until these railroads will consider it just as profitable to operate the other arm of their service with compound locomotives as the arm that they are now operating.

PROFESSOR ALLEN. Mr. Chairman, I would like to ask the speaker as to the increase in the proportion of compound engines built at their works last year, as compared with four or five years ago; to what extent has the use of the compound been increasing?

MR. VAUCLAIN. The gentleman has asked a question which is rather difficult to answer. I will, however, endeavor to do so. Four or five years ago the percentage of compound locomotives built by us, based on the entire output of the works, was about 22 per cent. at that particular time. It struck as high

BALDWIN "ATLANTIC" TYPE LOCOMOTIVE, P. & R. R. R.

BALDWIN LOCOMOTIVE WORKS

as 27 per cent. twice, but it fell down to about 22 per cent. Last year the percentage was about 35 per cent. of the total engines built, but this basis for estimating the percentage is erroneous, for this reason, that we must first eliminate switching engines, mining engines, electric locomotives, compressed air locomotives, and various other specialties which we build. After we have eliminated all of these we have left the road engines, and from these road engines we must eliminate the foreign locomotives which we build. The percentage of compound locomotives is then nearly 70 per cent. This is the fairest basis for one to form an idea as to the extent which compound locomotives are being introduced on American railroads at the present time.

THE CHAIRMAN. Mr. Vauclain said that he was willing to reply to any questions that might be asked him, and I think you will agree with me that his replies have been very instructive, and I am sure that we fully appreciate them. If there are no further questions we will consider the discussion closed.

BALDWIN LOCOMOTIVE WITH VANDERBILT BOILER

BALDWIN LOCOMOTIVE WORKS

Lehigh Valley Heavy Freight Locomotive

Weight on drivers, 202,232 pounds   Total weight, 225,082 pounds   Heating surface, 4,105 sq. feet

Baltimore and Ohio Heavy Freight Locomotive

Weight on drivers, 166,954 pounds   Total weight, 186,504 pounds   Heating surface, 2,354 sq. feet

BALDWIN LOCOMOTIVE WORKS

Union Pacific Heavy Freight Vanderbilt Boiler Locomotive

Weight on drivers, 174,000 pounds   Total weight, 196;000 pounds   Heating surface, 2,629 sq. feet

Pennsylvania R. R. High Speed "Atlantic" Type Locomotive

Heating surface, 2,278 sq. feet.   This is one of the finest specimens of "Atlantic" types in existence

BALDWIN LOCOMOTIVE WORKS

Chicago, Milwaukee and St. Paul Heavy Freight Locomotive, with 68-Inch Driving Wheels

Weight on drivers, 123,275 pounds   Total weight, 166,775 pounds   Heating surface, 2,715 sq. feet

Soo Line Heavy Freight Decapod Locomotive

Weight on drivers, 184,360 pounds   Total weight, 207,210 pounds   Heating surface, 3,015 sq. feet

BALDWIN LOCOMOTIVE WORKS

New York Central and Hudson River Heavy Ten-Wheel Passenger Locomotive

Weight on drivers, 134,205 pounds   Total weight, 175,005 pounds   Heating surface, 2,915 sq. feet

Baltimore and Ohio Heavy "Atlantic" Type Locomotive

Weight on drivers, 83,400 pounds   Total weight, 149,600 pounds   Heating surface, 2,663 sq. feet

BALDWIN LOCOMOTIVE WORKS

Chesapeake and Ohio Heavy Ten-Wheel Passenger Locomotive

Weight on drivers, 126,220 pounds   Total weight, 173,020 pounds   Heating surface, 3,000 sq. feet

Cambria Steel Company Heavy Shifting Locomotive

Weight on drivers, 137,080 pounds   Total weight, 137,080 pounds   Heating surface, 1,908 sq. feet

BALDWIN LOCOMOTIVE WORKS

Norfolk and Western Two-Cylinder Compound Freight Locomotive

Weight on drivers, 165,585 pounds   Total weight, 185,685 pounds   Heating surface, 2,788 sq. feet

Chicago, Burlington and Quincy High Speed "Atlantic" Type Compound Locomotive

Weight on drivers, 85,850 pounds   Total weight, 159,050 pounds   Heating surface, 2,510 sq. feet

BALDWIN LOCOMOTIVE WORKS

Chicago, Milwaukee and St. Paul High Speed "Atlantic" Type
Compound Locomotive

Weight on drivers, 71,600 pounds   Total weight, 140,700 pounds   Heating surface, 2,232 sq. feet

CORCOVADO RAILWAY, RIO DE JANEIRO, BRAZIL

Baldwin Rack Locomotive ascending grade of 25 per cent. crossing Silvestre Bridge.   The city and bay of Rio de Janeiro in the distance.

165

BALDWIN LOCOMOTIVE WORKS

Compound Atlantic Type Locomotive

for the

Class 10 $\frac{21}{40}$ ¼ -C-6

Gauge 4' 8½''

Canada Atlantic Railway Company

GENERAL DIMENSIONS

CYLINDERS

Diameter (High Pressure) .	13½''
" (Low Pressure) . .	23''
Stroke	26''
Valve . . . Balanced Piston	

BOILER

Diameter	62''
Thickness of Sheets . .	11/16''
Working Pressure . .	210 lbs.
Fuel	Soft Coal

FIREBOX

Material	Steel
Length	120⅛''
Width	40⅜''
Depth, Front . . .	74½''
" Back . . .	70½''
Thickness of Sheets, Sides .	⅜''
" " " Back .	⅜''
" " " Crown .	⅜''
" " " Tube .	½''

TUBES

Material	Iron
Number	258
Diameter	2''
Length	16' 0''

HEATING SURFACE

Firebox . .	186 sq. ft.
Tubes . . .	2150 sq. ft.
Total . . .	2336 sq. ft.
Grate Area . .	33.6 sq. ft.

DRIVING WHEELS

Diameter Outside . . .	84¼''
" of Center . .	78''
Journals . .	8½'' x 12''

ENGINE TRUCK WHEELS

Diameter . . .	36''
Journals . .	5½'' x 10''

TRAILING WHEELS

Diameter . . .	54¼''
Journals . .	8½'' x 12''

WHEEL BASE

Driving . . .	7' 6''
Rigid . . .	15' 0''
Total Engine . .	27' 1''
Total Engine and Tender .	54' 5½''

WEIGHT

On Driving Wheels .	86,030 lbs.
On Truck . . .	45,780 lbs.
" Trailing Wheels	37,100 lbs.
Total Engine .	168,910 lbs.
Total Engine and Tender	307,000 lbs.

TENDER

Diameter of Wheels .	36''
Journals . .	5½'' x 10''
Tank Capacity .	7,000 gals.

SERVICE

Fast Passenger.

BALDWIN LOCOMOTIVE WORKS

Atlantic Type Locomotive

for the

Long Island Railroad Company

Class 10-32¼-C-37

Gauge 4' 8½"

GENERAL DIMENSIONS

CYLINDERS

Diameter	19½"
Stroke	26"
Valve	Balanced

BOILER

Diameter	64"
Thickness of Sheets . .	11/16"
Working Pressure . .	200 lbs.
Fuel, coal: Anthracite, bituminous or mixed.	

FIREBOX

Material	Steel
Length	113¼"
Width	96"
Depth, Front . . .	60½"
" Back . . .	52½"
Thickness of Sheets, Sides .	⅜"
" " " Back .	⅜"
" " " Crown .	⅜"
" " " Tube .	½"

TUBES

Material	Iron
Number	297
Diameter	2"
Length	16' 0"

HEATING SURFACE

Firebox . . .	180.5 sq. ft.
Tubes . . .	2476.8 sq. ft.
Total . . .	2657.3 sq. ft.
Grate Area . .	75.5 sq. ft.

DRIVING WHEELS

Diameter Outside . . .	76"
" of Center . .	70"
Journals . . .	9" x 12"

ENGINE TRUCK WHEELS

Diameter . . .	36"
Journals . . .	5½" x 10"

TRAILING WHEELS

Diameter . . .	56"
Journals . . .	8" x 12"

WHEEL BASE

Driving	6' 7"
Rigid	13' 1"
Total Engine . . .	25' 1"
Total Engine and Tender .	53' 2¾"

WEIGHT

On Driving Wheels . .	97,850 lbs.
On Truck . . .	35,060 lbs.
" " Trailing Wheels	33,200 lbs.
Total Engine . .	166,110 lbs.
Total Engine and Tender	266,000 lbs.

TENDER

Diameter of Wheels . .	33"
Journals . . .	5" x 9"
Tank Capacity . .	5,000 gals.

SERVICE

Fast Passenger.

BALDWIN LOCOMOTIVE WORKS

American Type Locomotive

for the

Dominion Atlantic Railway Company

Class 8-30-C-558

Gauge 4′ 8½″

GENERAL DIMENSIONS

CYLINDERS

Diameter	18″
Stroke	24″
Valve	Balanced

BOILER

Diameter	60″
Thickness of Sheets .	⅝″ and 11⁄16″
Working Pressure . .	180 lbs.
Fuel	Soft Coal

FIREBOX

Material	Steel
Length	73 15⁄16″
Width	34⅜″
Depth, Front . . .	82¼″
" Back	80″
Thickness of Sheets, Sides .	5⁄16″
" " " Back .	5⁄16″
" " " Crown .	⅜″
" " " Tubes .	½″

TUBES

Material	Steel
Number	256
Diameter	2″
Length	10′ 11¼″

HEATING SURFACE

Firebox	144.5 sq. ft.
Tubes	1478.8 sq. ft.
Total	1623.3 sq. ft.
Grate Area . . .	17.9 sq. ft.

DRIVING WHEELS

Diameter Outside . . .	66″
" of Center . .	60″
Journals	8″ x 8½″

ENGINE TRUCK WHEELS

Diameter	30″
Journals	5″ x 10″

WHEEL BASE

Driving	8′ 9″
Total Engine . . .	22′ 9″
Total Engine and Tender .	45′ 8½″

WEIGHT

On Driving Wheels . .	66,210 lbs.
On Truck . . .	40,770 lbs.
Total Engine . . .	106,980 lbs.
Total Engine and Tender	182,000 lbs.

TENDER

Diameter of Wheels . .	36″
Journals	3¾″ x 7″
Tank Capacity . . .	3,750 gals.

SERVICE

Passenger.

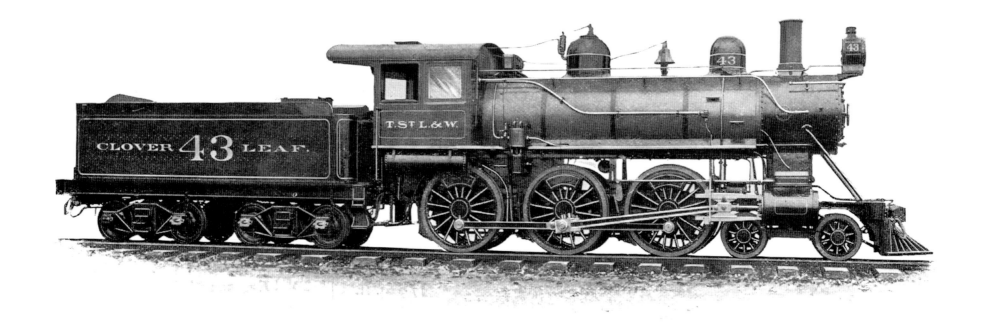

BALDWIN LOCOMOTIVE WORKS

Ten-Wheel Locomotive

Class 10-32-D-699

for the

Gauge 4' 8½"

Toledo, St. Louis and Western Railroad Company

GENERAL DIMENSIONS

CYLINDERS

Diameter	19"
Stroke	24"
Valve	Balanced

BOILER

Diameter	58"
Thickness of Sheets	9/16"
Working Pressure	180 lbs.
Fuel	Soft Coal

FIREBOX

Material	Steel
Length	96⅛"
Width	41"
Depth, Front	64½"
" Back	63"
Thickness of Sheets, Sides	⅜"
" " " Back	⅜"
" " " Crown	⅜"
" " " Tube	½"

TUBES

Material	Steel
Number	218
Diameter	2"
Length	13' 6"

HEATING SURFACE

Firebox	205.2 sq. ft.
Tubes	1531.4 sq. ft.
Total	1736.6 sq. ft.
Grate Area	27.36 sq. ft.

DRIVING WHEELS

Diameter Outside	68"
" of Center	62"
Journals	8" x 10"

ENGINE TRUCK WHEELS

Diameter	33"
Journals	5½" x 9"

WHEEL BASE

Driving	14' 0"
Total Engine	24' 4"
Total Engine and Tender	50' 8½"

WEIGHT

On Driving Wheels	94,850 lbs.
On Truck	33,335 lbs.
Total Engine	128,185 lbs.
Total Engine and Tender	228,000 lbs.

TENDER

Diameter of Wheels	36"
Journals	5" x 9"
Capacity of Tank	5,000 gals.

SERVICE

Passenger.

BALDWIN LOCOMOTIVE WORKS

Compound Ten-Wheel Locomotive

Class 10 $\frac{25}{46}$-D-63

for the

Gauge 4' 8½''

Union Pacific Railroad Company

GENERAL DIMENSIONS

CYLINDERS

Diameter (High Pressure) .	15½''
" (Low Pressure) . .	26''
Stroke	28''
Valve	Balanced Piston

BOILER

Diameter	66''
Thickness of Sheets .	11/16'' and ¾''
Working Pressure . .	200 lbs.
Fuel	Coal

FIREBOX

Material	Steel
Length	118 3/16''
Width	39⅛''
Depth, Front . . .	79½''
" Back . . .	67''
Thickness of Sheets, Sides .	5/16''
" " " Back .	5/16''
" " " Crown .	⅜''
" " " Tube .	½''

TUBES

Material	Iron
Number	350
Diameter	2''
Length	15' 6''

HEATING SURFACE

Firebox	186 sq. ft.
Tubes	2825 sq. ft.
Total	3011 sq. ft.
Grate Area . . .	32 sq. ft.

DRIVING WHEELS

Diameter Outside . . .	78''
" of Center . .	72''
Journals . . .	9'' x 12''

ENGINE TRUCK WHEELS

Diameter	30''
Journals . . .	6½'' x 11''

WHEEL BASE

Driving	14' 6''
Total Engine . . .	26' 5''
Total Engine and Tender .	54' 0''

WEIGHT

On Driving Wheels . .	141,320 lbs.
On Truck . . .	41,620 lbs.
Total Engine . .	182,940 lbs.
Total Engine and Tender	300,000 lbs.

TENDER

Diameter of Wheels . .	33''
Journals	5'' x 9''
Tank Capacity . .	6,000 gals.

SERVICE

Passenger.

BALDWIN LOCOMOTIVE WORKS

Compound Prairie Type Locomotive

Class 10 $\frac{26}{48}$ ¼-D-6

for the

Gauge 4′ 8½″

Chicago, Burlington & Quincy Railroad Company

GENERAL DIMENSIONS

CYLINDERS

Diameter (High Pressure) .	16″
" (Low Pressure) . .	27″
Stroke	24″
Valve . . .	Balanced Piston

BOILER

Diameter	65¼″
Thickness of Sheets . .	11/16″
Working Pressure . .	200 lbs.
Fuel . . .	Lignite Coal

FIREBOX

Material	Steel
Length	84″
Width	72″
Depth, Front . .	66¼″
" Back . .	63¼″
Thickness of Sheets, Sides .	⅜″
" " " Back .	⅜″
" " " Crown .	⅜″
" " " Tube .	½″

TUBES

Material	Steel
Number	272
Diameter . . .	2¼″
Length . . .	17′ 1¹¹/₁₆″

HEATING SURFACE

Firebox . . .	155.8 sq. ft.
Tubes . . .	2732.7 sq. ft.
Total . . .	2888.5 sq. ft.
Grate Area . .	42.0 sq. ft.

DRIVING WHEELS

Diameter Outside . .	64″
" of Center .	56″
Journals . . .	9″ x 10″

ENGINE TRUCK WHEELS

Diameter . . .	37¼″
Journals . . .	5½″ x 9″

TRAILING WHEELS

Diameter . . .	37″
Journals . .	6″ x 10″

WHEEL BASE

Driving . . .	12′ 1″
Total Engine . .	28′ 1″
Total Engine and Tender .	54′ 6½″

WEIGHT

On Driving Wheels . .	132,838 lbs.
On Truck, Front . .	17,034 lbs.
" " Trailing Wheels	26,320 lbs.
Total Engine . .	176,192 lbs.
Total Engine and Tender .	296,000 lbs.

TENDER

Diameter of Wheels . .	36″
Journals . . .	5″ x 9″
Tank Capacity . .	6,000 gals.

SERVICE

Freight.

BALDWIN LOCOMOTIVE WORKS

Mogul Locomotive

Class 8-34-D-90

for the

Gauge 4′ 8½″

Pennsylvania Railroad Company

GENERAL DIMENSIONS

CYLINDERS

Diameter	20″
Stroke	28″
Valve	Balanced

BOILER

Diameter	68¾″
Thickness of Sheets	¾″
Working Pressure	205 lbs.
Fuel	Soft Coal

FIREBOX

Material	Steel
Length	111³⁄₁₆″
Width	39⅞″
Depth, Front	77⅝″
" Back	65⅝″
Thickness of Sheets, Sides	⁵⁄₁₆″
" " " Back	⁵⁄₁₆″
" " " Crown	⅜″
" " " Tube	½″

TUBES

Material	Iron
Number	356
Diameter	2″
Length	12′ 2¾″

HEATING SURFACE

Firebox	166.5 sq. ft.
Tubes	2264.8 sq. ft.
Total	2431.3 sq. ft.
Grate Area	30.25 sq. ft.

DRIVING WHEELS

Diameter Outside	62″
" of Center	55″
Journals, Main	9¼″ x 12″
" Front	8½″ x 12″
" Back	8½″ x 12″

ENGINE TRUCK WHEELS

Diameter	33″
Journals	5½″ x 10″

WHEEL BASE

Driving	14′ 9″
Total Engine	23′ 10″
Total Engine and Tender	54′ 5″

WEIGHT

On Driving Wheels	139,100 lbs.
On Truck	20,900 lbs.
Total Engine	160,000 lbs.
Total Engine and Tender	280,000 lbs.

TENDER

Diameter of Wheels	33″
Journals	5″ x 9″
Tank Capacity	6,000 gals.

SERVICE

Freight.

BALDWIN LOCOMOTIVE WORKS

Compound Consolidation Locomotive

Class 10 $\frac{28}{50}$-E-65

for the

Chicago Great Western Railway Company

Gauge 4′ 8½″

GENERAL DIMENSIONS

CYLINDERS

Diameter (High Pressure) .	17″
" (Low Pressure) . .	28″
Stroke	30″
Valve . . .	Balanced Piston

BOILER

Diameter	68″
Thickness of Sheets .	11/16 and ¾″
Working Pressure . .	200 lbs.
Fuel	Soft Coal

FIREBOX

Material	Steel
Length	108 3/16″
Width	42″
Depth, Front . . .	75″
" Back . . .	72″
Thickness of Sheets, Sides .	5/16″
" " " Back .	5/16″
" " " Crown .	3/8″
" " " Tube .	½″

TUBES

Material	Iron
Number	321
Diameter	2″
Length	13′ 6″

HEATING SURFACE

Firebox	182 sq. ft.
Tubes	2252 sq. ft.
Total	2434 sq. ft.
Grate Area . . .	31.5 sq. ft.

DRIVING WHEELS

Diameter Outside . . .	55″
" of Center . .	48″
Journals	9″ x 12″

ENGINE TRUCK WHEELS

Diameter	30″
Journals	6″ x 12″

WHEEL BASE

Driving	15′ 1″
Total Engine . . .	23′ 9″
Total Engine and Tender .	51′ 6½″

WEIGHT

On Driving Wheels . .	160,200 lbs.
On Truck . . .	18,200 lbs.
Total Engine . . .	178,400 lbs.
Total Engine and Tender .	297,000 lbs.

TENDER

Diameter of Wheels . .	33″
Journals	5″ x 9″
Tank Capacity . .	6,000 gals.

SERVICE

Freight.

Guaranteed to haul 1300 tons, exclusive of engine and tender, up a one per cent. grade 5 miles long, at a minimum speed of 10 miles per hour.

BALDWIN LOCOMOTIVE WORKS

Consolidation Locomotive

Class 10-38-E-341

for the

Gauge 4' 8½"

Chicago Terminal Transfer Railroad Company

GENERAL DIMENSIONS

CYLINDERS

Diameter	22"
Stroke	28"
Valve	Balanced

BOILER

Diameter	74"
Thickness of Sheets . . .	7/8"
Working Pressure . . .	220 lbs.
Fuel	Soft Coal

FIREBOX

Material	Steel
Length	102"
Width	66"
Depth, Front . . .	68½"
" Back . . .	63½"
Thickness of Sheets, Sides .	3/8"
" " " Back .	3/8"
" " " Crown .	3/8"
" " " Tube .	5/8"

TUBES

Material	Iron
Number	347
Diameter	2"
Length	14' 6"

HEATING SURFACE

Firebox . . .	172.5 sq. ft.
Tubes . . .	2613.6 sq. ft.
Total . . .	2786.1 sq. ft.
Grate Area	46.7 sq. ft.

DRIVING WHEELS

Diameter Outside . . .	51"
" of Center . .	44"
Journals . . .	9" x 13"

ENGINE TRUCK WHEELS

Diameter	30"
Journals . . .	5½" x 12"

WHEEL BASE

Driving	14' 8"
Total Engine . . .	23' 5"
Total Engine and Tender . .	53' 6"

WEIGHT

On Driving Wheels . .	169,140 lbs.
On Truck . . .	21,150 lbs.
Total Engine . . .	190,290 lbs.
Total Engine and Tender .	310,000 lbs.

TENDER

Diameter of Wheels . . .	33"
Journals	5" x 9"
Tank Capacity . .	6,000 gals.

SERVICE

Freight

Grades, one per cent.

BALDWIN LOCOMOTIVE WORKS

Consolidation Locomotive

Class 10-38-E-310

for the

Gauge 4' 8½"

Colorado Fuel and Iron Company

GENERAL DIMENSIONS

CYLINDERS

Diameter	22″
Stroke	28″
Valve	Balanced

BOILER

Diameter	76″
Thickness of Sheets . .	¾″
Working Pressure . .	170 lbs.
Fuel	Soft Coa

FIREBOX

Material	Steel
Length	125¹¹⁄₁₆″
Width	42⅝″
Depth, Front . . .	71½″
" Back . . .	67½″
Thickness of Sheets, Sides .	5⁄16″
" " " Back .	5⁄16″
" " " Crown .	⅜″
" " " Tube .	½″

TUBES

Material	Iron
Number	300
Diameter	2¼″
Length	13' 10″

HEATING SURFACE

Firebox	200 sq. ft.
Tubes	2430 sq. ft.
Total	2630 sq. ft.
Grate Area . . .	37 sq. ft.

DRIVING WHEELS

Diameter Outside . . .	50″
" of Center . . .	44″
Journals	9″ x 12″

ENGINE TRUCK WHEELS

Diameter	29¼″
Journals	5″ x 10″

WHEEL BASE

Driving	14' 3″
Total Engine . . .	22' 10″
Total Engine and Tender .	51' 1½″

WEIGHT

On Driving Wheels .	155,200 lbs.
On Truck . . .	14,850 lbs.
Total Engine . . .	170,050 lbs.
Total Engine and Tender .	250,000 lbs.

TENDER

Diameter of Wheels . .	30″
Journals	4¼″ x 8″
Tank Capacity . .	4,000 gals.

SERVICE

Freight.

Curves 20 degrees radius. Grades up to three per cent. Estimated to haul 400 tons, cars and lading up a straight grade of three per cent.

COO-LEY KU-I-TAN

BALDWIN LOCOMOTIVE WORKS

Consolidation Locomotive

Class 10-34-E-1578

for the

Gauge 4' 8½"

Simpson Logging Company

GENERAL DIMENSIONS

CYLINDERS

Diameter	20"
Stroke	24"
Valve	Balanced

BOILER

Diameter	66"
Thickness of Sheets . .	11/16"
Working Pressure . .	180 lbs.
Fuel	Wood

FIREBOX

Material	Steel
Length	112³⁄₁₆"
Width	42⅝"
Depth, Front . . .	65½"
" Back . . .	61½"
Thickness of Sheets, Sides .	5/16"
" " " Back .	5/16"
" " " Crown .	3/8"
" " " Tube .	1/2"

TUBES

Material	Iron
Number	270
Diameter	2"
Length	12' 5"

HEATING SURFACE

Firebox	166 sq. ft.
Tubes	1744 sq. ft.
Total	1910 sq. ft.
Grate Area . . .	33.52 sq. ft.

DRIVING WHEELS

Diameter Outside . . .	50"
" of Center . .	44"
Journals	8" x 9"

ENGINE TRUCK WHEELS

Diameter	30"
Journals	5½" x 10"

WHEEL BASE

Driving	13' 6"
Total Engine . . .	21' 8"
Total Engine and Tender .	50' 3"

WEIGHT

On Driving Wheels .	115,445 lbs.
On Truck . . .	15,600 lbs.
Total Engine . .	131,045 lbs.
Total Engine and Tender .	187,000 lbs.

TENDER

Diameter of Wheels . .	30"
Journals	3¾" x 7"
Tank Capacity . . .	2,800 gals.

SERVICE

Logging.

Engine used to haul logs without the use of trucks or cars. For this purpose two pieces of 2" x 12" planking are fastened to the crossties between the rails, along the full length of the road. On this runway the logs are dragged from the camp to the point of shipment.

The total length of the road is 6 miles. The grade on the first section is 7 per cent., continuous for 1¼ miles. On the second section it varies from level to 3.25 per cent. The curves vary from 8 to 14 degrees, contouring the side of the hill with 50 to 150-foot tangents.

The logs are hauled down the grade at a speed of 20 miles per hour. On the up grade, with one standard freight car with 30 tons load, the speed on the first section is 6 to 8 miles per hour, and on the second section 20 miles per hour.

BALDWIN LOCOMOTIVE WORKS

American Type Locomotive

Class 8-22-C-119

for the

F. C. Santander á Bilbao (Spain)

Gauge 3' 3⅜"

GENERAL DIMENSIONS

CYLINDERS

Diameter	13½"
Stroke	24"
Valve	Balanced

BOILER

Diameter	46"
Thickness of Sheets . . .	½"
Working Pressure . . .	160 lbs.
Fuel	Soft Coal

FIREBOX

Material	Copper
Length	70"
Width	25"
Depth, Front . . .	46¾"
" Back . .	37½"
Thickness of Sheets, Sides .	½"
" " " Back . .	½"
" " " Crown .	½"
" " " Tube	¾" and ½"

TUBES

Material	Brass
Number	118
Diameter . . .	2"
Length . . .	10' 7¾"

HEATING SURFACE

Firebox . . .	66.2 sq. ft.
Tubes . . .	650.8 sq. ft.
Total . . .	717 sq. ft.
Grate Area . .	12.15 sq. ft.

DRIVING WHEELS

Diameter of Outside . .	51½"
" of Center . .	46"
Journals . . .	6" x 7"

ENGINE TRUCK WHEELS

Diameter . . .	24"
Journals . . .	4" x 6½"

WHEEL BASE

Driving . . .	6' 6"
Total . . .	19' 6"

WEIGHT

On Driving Wheels . .	52,155 lbs.
On Truck . .	22,015 lbs.
Total . . .	74,170 lbs.

TANK

Tank Capacity . . .	740 gals.

SERVICE

Passenger

BALDWIN LOCOMOTIVE WORKS

Six-Coupled Double Ender Locomotive

Class 10-26¼-D-14

for the

Gauge 4' 8½''

Seoul & Chemulpo Railway (Korea)

GENERAL DIMENSIONS

CYLINDERS

Diameter	16''
Stroke	24''
Valve	Balanced

BOILER

Diameter	52''
Thickness of Sheets . .	½''
Working Pressure . .	160 lbs.
Fuel	Soft Coal

FIREBOX

Material	Steel
Length	78⁷⁄₁₆''
Width	33⅜''
Depth, Front . . .	56½''
'' Back . . .	41''
Thickness of Sheets, Sides .	⁵⁄₁₆''
'' '' '' Back .	⁵⁄₁₆''
'' '' '' Crown .	⅜''
'' '' '' Tube .	½''

TUBES

Material	Iron
Number	176
Diameter	2''
Length . . .	10' 6''

HEATING SURFACE

Firebox . . .	107 sq. ft.
Tubes	950 sq. ft.
Total . . .	1057 sq. ft.
Grate Area . . .	18.2 sq. ft.

DRIVING WHEELS

Diameter Outside . . .	54''
'' of Center . . .	48''
Journals . . .	6½'' x 8''

ENGINE TRUCK WHEELS (Front)

Diameter	29½''
Journals . . .	4¼'' x 7''

ENGINE TRUCK WHEELS (Back)

Diameter	29½''
Journals . . .	4¼'' x 7''

WHEEL BASE

Driving	11' 4''
Total	25' 2''

WEIGHT

On Driving Wheels . .	79,855 lbs.
On Truck, Front . .	16,300 lbs.
'' '' Back .	14,350 lbs.
Total	110,505 lbs.

TANK

Tank Capacity . . .	1,500 gals.

SERVICE

Passenger and freight.

Road 26¼ miles in length. Curves 6 degrees. Grade one per cent. Engine to haul ten to twelve cars, weighing 50,000 pounds each, up a straight grade of one per cent., track and cars in good condition. Average speed 22 miles per hour.

BALDWIN LOCOMOTIVE WORKS

Compound Rack Locomotive

for the

Manitou & Pike's Peak Railway

Class 6 $\frac{14}{24}$ $\frac{1}{3}$-C-5

Gauge 4' 8⅝"

GENERAL DIMENSIONS

CYLINDERS

Diameter (High Pressure) .	10"
" (Low Pressure) . .	15"
Stroke	22"
Valve . . .	Balanced Piston

BOILER

Diameter	44"
Thickness of Sheets . .	7/16"
Working Pressure . .	180 lbs.
Fuel	Hard Coal

FIREBOX

Material	Steel
Length	48"
Width	59⅜"
Depth, Front . . .	46¼"
" Back . . .	40⅝"
Thickness of Sheets, Sides .	5/16"
" " " Back .	5/16"
" " " Crown .	⅜"
" " " Tube .	7/16"

TUBES

Material	Steel
Number	176
Diameter	1½"
Length	7' 5¹⁵/₁₆"

HEATING SURFACE

Firebox . .	58.25 sq. ft.
Tubes . . .	518.25 sq. ft.
Total . . .	576.50 sq. ft.
Grate Area . .	19.79 sq. ft.

DRIVING WHEELS

Carrying Wheels, Dia. Outside	25¾"
" Dia. of Center	21"
Rack Wheel Pitch, Diameter	22.468"
Journals	6" x 6"

ENGINE TRUCK WHEELS

Diameter	25½"
Journals	4" x 6"

WHEEL BASE

Driving	5' 7"
Total	12' 3"

WEIGHT

On Driving Wheels . .	44,155 lbs.
On Truck . . .	18,300 lbs.
Total . . .	62,455 lbs.

TANK

Tank Capacity . .	600 gals.

SERVICE

Passenger.

"Abt" System of Rack Rail.

Grade 25 per cent.

LOCOMOTIVE, CLASS F-3-B, WITH BROAD FIREBOX, FOR THE PENNSYLVANIA RAILROAD COMPANY

Broad Firebox Locomotives

A Paper read by Samuel M. Vauclain, before the

Pennsylvania Railroad Y. M. C. A.

Philadelphia, March 25, 1901

MR. CHAIRMAN AND GENTLEMEN:—It is, I assure you, with great pleasure that I address you to-night. The subject you have chosen for my consideration and your enlightenment, namely: "The Wootten or Broad Firebox Boiler," is one with which I am quite familiar, therefore an easy task is before me.

To those of you not aware of the fact, let me say that the so-called Wootten boiler was first invented and brought out by John E. Wootten, then Superintendent of Motive Power of the Philadelphia and Reading Railway Company, about the year 1877, and while much credit is due him for its conception, and his persistence in continuing the experiments towards its final introduction, the mechanical world is chiefly indebted to L. B. Paxson, formerly Superintendent of Motive Power of the Philadelphia and Reading Railway Company, and now retired, for the great success that has attended its career. Mr. Paxson's engineering skill, untiring energy and continuity of purpose, beyond a doubt caused the success of the Wootten boiler for locomotives, a boiler that to-day is being modified in almost every conceivable shape, and under fictitious names, representing the principles on which the Wootten boiler was based.

BALDWIN LOCOMOTIVE WORKS

Now, while Mr. Paxson was busily engaged developing the genuine Wootten boiler, a rival appeared on the scene, in no less a person than Alexander Mitchell, of the Lehigh Valley Railroad Company, known to all locomotive people as the father of the Consolidation locomotive. The first of these locomotives was built for him at the Baldwin Locomotive Works, in 1866, and to-day thousands are in use the world over. Mr. Mitchell's contention was that the combustion chamber, which was considered so essential by Mr. Wootten and Mr. Paxson, was an unnecessary and undesirable adjunct and would cause trouble; whereas with a straight flue sheet greater heating surface would be obtained and the brick wall in the combustion chamber would not be necessary. To more satisfactorily explain to you the theories of these respective rivals, I will first explain the internal economy of the Wootten boiler, from its origin until the present time.

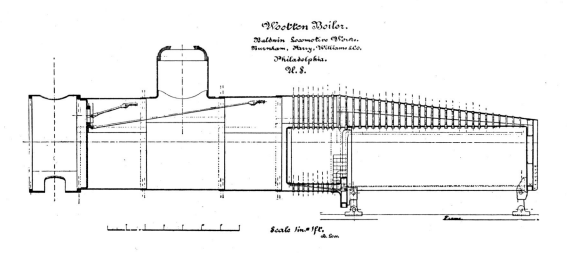

FIGURE 1

Figure 1, illustrated herewith, shows how the first Wootten boilers were constructed. You will notice that in addition to the large overhanging but very shallow firebox, a combustion chamber was projected into the barrel or waist of the boiler. A water leg was provided next the throat sheet, which also extended above the bottom of the combustion chamber, and on which the bridge wall was built. The mud ring or water space frame was made out of U-shaped metal, bent and ingeniously joined at the corners. It was found

BALDWIN LOCOMOTIVE WORKS

that while this arrangement burned refuse fuel successfully, the boiler gave trouble from many causes, chief among the number being the flanged firebox frame and the breakage of staybolts in the space immediately over the fire and between the sides and crown sheet. The irregular angle at which they penetrated the sheets, and the angularity of the contour of the firebox were also soon found objectionable.

Figure 2, illustrated herewith, is a cross section of the boiler, which shows the irregular staying and the sharp corners of the firebox roof or crown sheet.

Figure 3, shown on page 198, is the next marked step in design. The sloping top sheet has disappeared and a solid water-space frame of rectangular section holds the firebox securely at its base. An effort has been made to more evenly space the staybolts, but the water space under the combustion chamber wall is still retained.

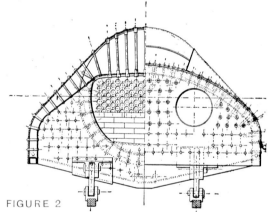

FIGURE 2

You will notice that this boiler is a wagon-top boiler, also that the stays in the crown sheet are radial thereto, and that the dome is on the extended portion of the wagon top. It was this boiler, patented by Mr. Wootten, which prevented William L. Austin, of the Baldwin Locomotive Works, from obtaining letters patent on what is commonly called to-day the Radial Stayed Extended Wagon-top Boiler, thousands of which are running, and owe their existence to his inventive and engineering genius.

Wootten boilers of this type, however, still gave trouble, not as much as before, but quite enough to cause many railway people to keep hands off. About this time the Baldwin Locomotive Works took an

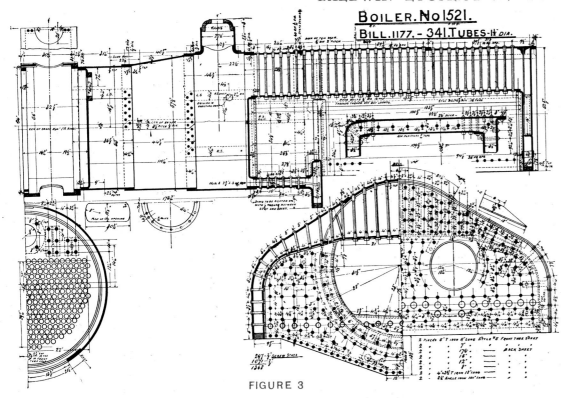

BALDWIN LOCOMOTIVE WORKS

BOILER. No 1521.

BILL. 1177. - 341.TUBES-1⅛ DIA.

FIGURE 3

active interest in perfecting these boilers, and as soon as they were permitted to put the boiler into good shape, a demand was created.

Figure 4, shown on page 199, is the next step toward perfection. The crown is made circular until it reaches the side sheets, the staybolts are all radial to its surface, the water leg in the combustion chamber has disappeared, and the crown sheet and combustion chamber-roof sheets are in one plate, thus omitting the seam at their junction which always gave trouble.

Figure 5, shown on page 200, gives you the latest and best shape of the regular Wootten boiler, used on the ten-wheel locomotives of the Philadelphia and Reading Railway. The firebox is almost semi-circular, and the outer shell sheets are curved at every point and somewhat semi-elliptical; the combustion chamber is flattened at the bottom in order to make space in the throat sheet for the water tubes, as the firebox frame is usually on a line with the bottom of the boiler shell, and no space would otherwise be available.

With this explanation I will now turn to the ideal of Mr. Mitchell, or what is known as a modified Wootten, the modification being to omit the combustion chamber.

BALDWIN LOCOMOTIVE WORKS

Figure 6, shown on page 201, is Mr. Mitchell's original straight flue sheet boiler. The resemblance to the previous illustrations is marked, but you will observe that the flue sheet is perfectly straight. In practice this feature gave infinitely more trouble than the combustion chamber, due no doubt to the great amount of cold air which at times came through the shallow fire and passed over the ends of and through the tubes.

Many attempts to remedy this defect were tried from time to time, but the final result was a return to the idea of a combustion chamber, very much shortened, however, and called a D head. This placed the tube sheet about six inches ahead of the throat sheet, and prevented the cold air striking the flue ends, and thus causing them to leak; it also prevented the flue ends from coming in direct contact with the fire.

Figure 7, shown on page 202, is the most striking example of a boiler of this modified type, and is the result of the fearless engineering of Samuel Higgins, Superintendent of Motive Power of the Lehigh Valley Railroad Company, who followed in Mr. Mitchell's footsteps. This boiler has 511 tubes, two inches in diameter, and over 4000 feet of heating surface.

Figure 8, shown on page 203, is a cross section of the boiler just mentioned; from it can be had a better idea of the arrangement of the staybolts and the con-

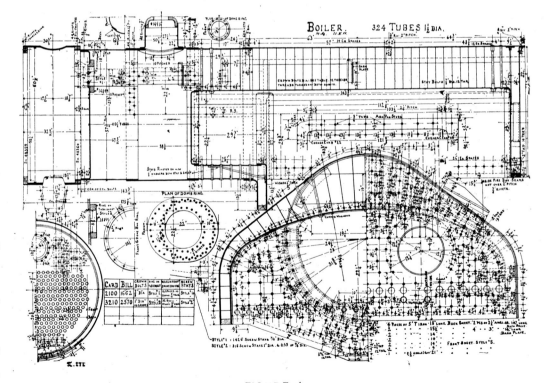

FIGURE 4

tour of the sheets, also a fair idea of what the fireman has to do to keep the firebox filled; but fitted as it is to a Baldwin Compound locomotive, one man has no difficulty in keeping the safety valves blowing when the engine is doing the work of two ordinary locomotives. The engine referred to, of which there are several on the mountains near Wilkesbarre, is shown in Figure 9, page 204.

This locomotive weighs 230,000 pounds in working order, has a maximum tractive effort of 54,000 pounds, and is compounded in the ordinary Baldwin manner.

FIGURE 5

Figure 10, shown on page 204, gives an idea of the early Wootten locomotive fitted with the first boiler described. Its general appearance would cause unfavorable comment to-day, and must be considered merely as representing the art twenty years ago.

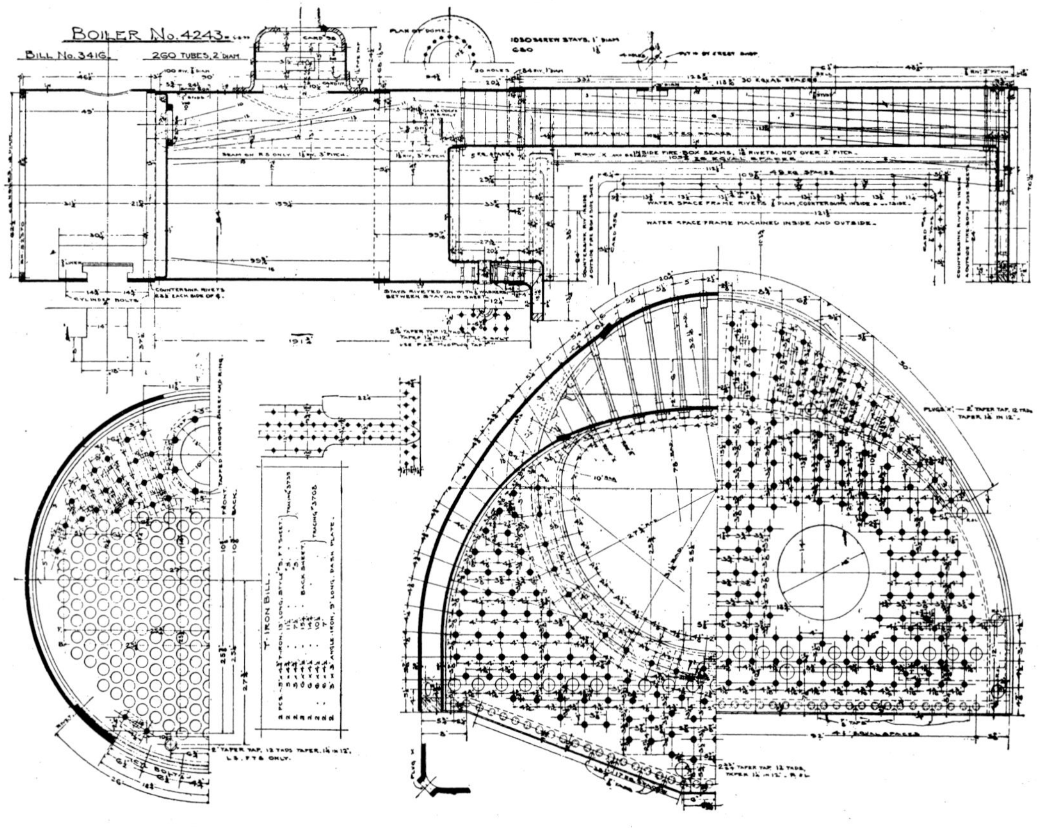

Figure 11, shown on page 205, is a locomotive of similar class, fitted with a boiler of later design. These locomotives gave us all, I assure you, some lessons on high speed. With less than 1400 square feet of heating surface, and only a sixty-eight-inch wheel, they held the record for high speed during their era, and hauled all the fast two-hour trains to New York, demonstrating that without large grates heating surface was of no use. The speaker's experience with these engines made him an early convert to the Wootten boiler, and its proper application has since been a constant study.

Figure 12, on page 206, shows one of the celebrated Black Diamond engines of the Lehigh Valley Railroad, equipped with modified Wootten boiler. They have made themselves world famous, not only for high speed, but for their regularity of arrival at terminals. On a recent visit to Rochester, it was noted that of all the roads centering there, this was the only

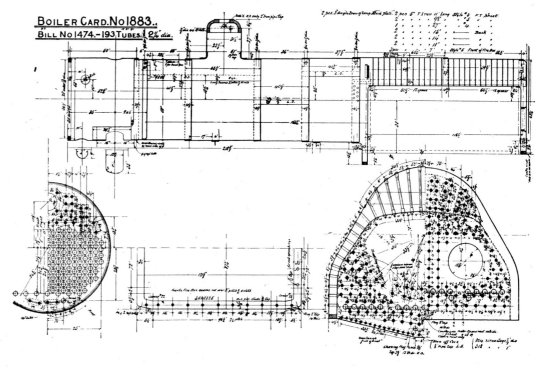

FIGURE 6

engine to bring its train in on time—a Wootten firebox victory. The next consignment of these engines, now building, will be superior to those at present running, as compound cylinders of the Baldwin four-cylinder type are to be applied to them.

Figure 13, on page 207, shows No. 1027. This locomotive (a compound by the way, with only 1800 square feet of heating surface,) made itself world-famous. Many of the leading mechanics of Europe were excited over its remarkable performance on one of our seashore lines. Fitted with a combustion chamber, and burning anthracite coal, egg size, this engine covered the fifty-five miles every day throughout the season, in less than

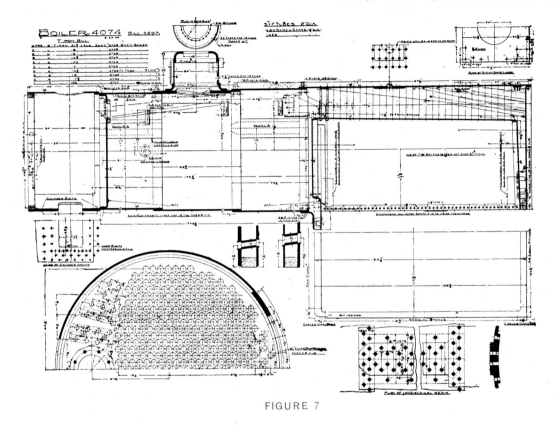

FIGURE 7

fifty minutes; many days in forty-five, and seldom over forty-eight. Those of you who sit at the throttle know what this means much better than those who do their railroading riding in a Pullman car, and can appreciate it accordingly. Larger and more powerful engines of this class have been made, but none have a more satisfactory record.

Let us now consider the capacity of the Wootten type of boiler, taking into consideration the great advantage obtained by its use in developing large horse-power.

It has been definitely proved, and is an accepted fact, that for anthracite coal, both refuse and selected, no boiler has given such good results as are being daily obtained from the Wootten boiler. Many railroads have, with commendable tenacity, adhered to a forty-two-inch wide firebox, ten feet long, and even eleven feet long, in a vain effort to equal, if not exceed, the results obtained from Wootten boilers. Many mechanical engineers, however, have at last accepted the principles upon which the Wootten boiler is constructed, and the result is a world-wide revolution in locomotive boiler design.

BALDWIN LOCOMOTIVE WORKS

In order to handle our present heavy trains at high speeds, large steaming capacity is necessary, and as the horse-power of a locomotive increases in proportion to the tractive effort multiplied by the speed, it is evident to you that if a locomotive with 1500 square feet of heating surface can haul a given train thirty miles per hour, that to haul it sixty miles per hour would require more than double the heating surface, the usual allowance being made for the increased tractive effort at the higher speed. Now to do this satisfactorily, would require a boiler of more than double the heating surface, and an equivalent increase in grate surface as well, as it would not be possible to burn with economy more than twice the coal on the small grate as used in the smaller boiler; consequently, if boilers having 1500 square feet of heating surface and twenty-five square feet of grate surface were satisfactory and efficient, it follows that if 3000 square feet of heating surface is necessary, twice twenty-five or fifty square feet of grate surface should be provided.

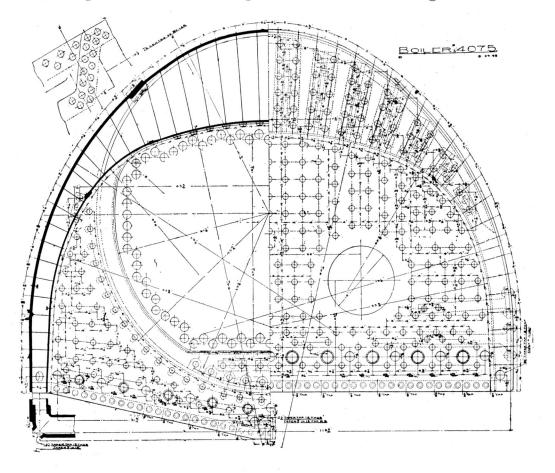

FIGURE 8

BALDWIN LOCOMOTIVE WORKS

FIGURE 9

It has also been definitely determined that the most economical ratio of grate surface to heating surface for bituminous coal is one to sixty, that is for each sixty square feet of heating surface in the boiler one square foot of grate surface should be supplied. For anthracite coal, a ratio of one to forty is required. It will thus be seen that in a boiler having 4000 square feet of heating surface, 100 square feet of grate surface should be supplied for anthracite coal, and sixty-six and two-thirds square feet for bituminous coal.

The average modern locomotive of the present day has about 3000 square feet of heating surface, requiring fifty square feet of grate surface, and the endeavor to so provide is to-day very marked, as can be seen by a visit to the Baldwin Locomotive Works, where any and all are welcome. The ability of the Wootten boiler, with its wide grate, to economically consume the

FIGURE 10

BALDWIN LOCOMOTIVE WORKS

FIGURE 11

fuel, not only insures rapid generation of steam to maintain high horse-power, but permits the use of larger exhaust nozzles, thus practically freeing the cylinders from back pressure, and adding to the effective work of the loco- motive. Engineers and firemen at first object to this boiler, as it promises more work; but they soon find that by the more perfect combustion of the fuel a much less quantity is necessary to do the work. As much economy can be effected by applying this large firebox as is ordinarily obtained by compounding; but when both are in use the combination is complete, and the fireman's work reduced forty per cent. in freight service.

The original object in view by Mr. Wootten, however, was to burn the refuse or unmarketable sizes of anthracite coal. In this he was very successful for freight work, but not until the application of compound cylinders was it possible to do this on high-speed passenger trains. To-day, however, all the fast New York trains on the Reading Railway are operated with this class of fuel, with compound Wootten engines, with a total saving of sixty-eight per cent. in the cost of fuel.

For several years the Baldwin Locomotive Works has not only offered, but has guaranteed, to build locomotives to burn successfully any fuel which may be sent to them, and in every case, after a careful analysis, some form of the Wootten boiler has been found necessary to enable them to successfully carry out their contracts.

BALDWIN LOCOMOTIVE WORKS

FIGURE 12

In 1895 it was suggested to the Japan Railway Company, of Japan, that if a sample of coal found in great quantities on the line of their road, and considered unfit for locomotive use (they having been unable to burn it in the English locomotives with small fireboxes), could be sent to the Baldwin Locomotive Works, the latter would build and guarantee locomotives to burn it. The fuel was sent. It came from two mines, the Iwaki mine and the Iryana mine. The analysis was as follows:

IWAKI MINE

Moisture	7.19 per cent.
Volatile matter	46.10 per cent.
Fixed carbon	32.39 per cent.
Ash	14.32 per cent.

Coal will not clinker.

IRYANA MINE

Moisture	10.78 per cent.
Volatile matter	39.49 per cent.
Fixed carbon	40.31 per cent.
Ash	9.42 per cent.

Coal will clinker.

BALDWIN LOCOMOTIVE WORKS

This coal was very difficult to ignite, on account of the great amount of moisture it contained; but after it had once been ignited it burned freely but not rapidly. The first coal, from the Iwaki mine, was composed of lumps about the size of the ordinary egg coal in the anthracite regions. The second coal, from the Iryana mine, was composed of lumps about the size of broken coal, a little larger than egg coal.

Figure 14, shown on page 208, was the type of locomotive constructed for this purpose, for passenger service. This locomotive was fitted with the modified type of Wootten boiler. The grate area was calculated to burn a sufficient amount of this fuel to generate steam enough for the tractive power desired, or the tractive power called for by the size of the cylinders. The reason for this was that the coal was peculiar. Taking the time required to ignite the coal and get it to a proper state of combustion, was about two or two and a half times longer than the ordinary Westmoreland gas coal, which, as you all know, is a quick and free burning coal.

FIGURE 13

Figure 15, shown on page 209, is one of the locomotives built for freight service of the Consolidation type. These are the largest Consolidation locomotives that have ever been shipped to Japan. This boiler is also of the modified Wootten type, large in capacity, and large in diameter. When you understand that we not only built one of each of these locomotives, but forty-four of them in one lot, guaranteed them to burn this fuel, and shipped them to a foreign country, you will believe that we had confidence in what we were doing, and that the locomotives would turn out all right. They gave excellent satisfaction, and are wonderfully free steamers. We were enabled to enlarge the exhausts, so that the engines were almost noiseless in their work.

In this country we have examples just as striking.

Figure 16, shown on page 210, is a locomotive built for the Missouri, Kansas and Texas Railway, and guaranteed to burn soft coal refuse from the mines along the line. The Missouri, Kansas and Texas, as you know, begins at St. Louis, passes down through Missouri and Kansas, and then through the Indian Territory into Texas. In the Indian Territory, at the railway company's various coal mines, much refuse or slack coal had accumulated. This coal was considered so inferior

FIGURE 14

that it did not even make good filling, and the mine owners were desirous of getting rid of it, and the railway thought that if it could be burned they might save some money. We agreed to construct a locomotive to do it, if they would send a sample of the fuel for analysis. The fuel was sent and the analysis of the coal is as follows:

Quality of coal	Screenings.
Moisture	2.70 per cent.
Volatile matter	33.22 per cent.
Fixed carbon	23.11 per cent.
Ash	40.88 per cent.

Theoretical evaporation, water per pound coal, 7.99. Coal will clinker. Size very fine.

BALDWIN LOCOMOTIVE WORKS

FIGURE 15

This presented a decidedly tough job. After going through the matter and figuring upon the value of the various elements in the fuel, it was decided to employ the regular Wootten type of firebox, that is, a firebox with a long combustion chamber. This was due principally to the fact that the coal was very fine, almost dust, and we were afraid that with the modified type of Wootten boiler the exhaust would simply take the dust off the shovel and carry it right through the tubes. Therefore we employed a high brick wall in the combustion chamber, and a low, open, single exhaust in a short smokebox in front. The result obtained by this locomotive can be better understood if you will permit me to read a letter from the Superintendent of Motive Power of that railroad.

MISSOURI, KANSAS AND TEXAS RAILWAY,

PARSONS, KAN., March 7, 1898.

MESSRS. BURNHAM, WILLIAMS & CO.,

BALDWIN LOCOMOTIVE WORKS,

PHILADELPHIA, PA.

GENTLEMEN:—If entirely convenient, you will kindly send me a copy of the analysis of slack coal from our mines, sample of which was sent you previous to the building of our slack-burning engine No. 251. The report sent to our management has been mislaid, or at least it is not now avail-able. I am in receipt of inquiries from several roads relative to efficiency of this type of boiler, and in order to give the engine what is justly due her, I would be pleased to give them the analysis of the coal, showing its inferior quality.

It will no doubt interest you to know that this engine shows a saving of fuel of $29.05 per round trip of 314 miles, compared with other engines of the same class, doing same work, burning run of mine coal. It is possible that we may interest other roads in the matter of slack-burning engines.

Very truly yours,   WILLIAM O'HERIN.

FIGURE 16

On April 20, 1899, the speaker read a paper before the New York Railroad Club, in which he drew the attention of railway managers to the enormous economies that were obtain-able by using some form of Wootten or wide firebox boiler. This paper was afterwards printed in full, with the discussion which followed, in No. 15 of a series of Recent Construction pam-phlets by the Baldwin Locomotive Works. Thousands of these have been

FIGURE 17

BALDWIN LOCOMOTIVE WORKS

FIGURE 18

distributed all over the world, and are still sought after from home and abroad.

Their distribution has resulted in a wide-spread demand for some form of wide or Wootten firebox, and most notable among those adopting this type of boiler are the following:

Figure 17, shown on page 210, is a Baltimore and Ohio Compound Wootten consolidation. The Baltimore and Ohio Railroad Company has about 135 of these engines, which in regular service show about ten per cent. saving in fuel over the narrow or forty-two-inch wide firebox compound engine of the same size, and about thirty-five per cent. saving over the narrow firebox simple engine of the same size.

Figure 18, illustrated herewith, is a Baltimore and Ohio Atlantic type locomotive. This is the most modern passenger engine on the Baltimore and Ohio Railroad, and has been only recently delivered to them; it has the wide firebox, and is a most excellent steamer and very easy to fire.

FIGURE 19

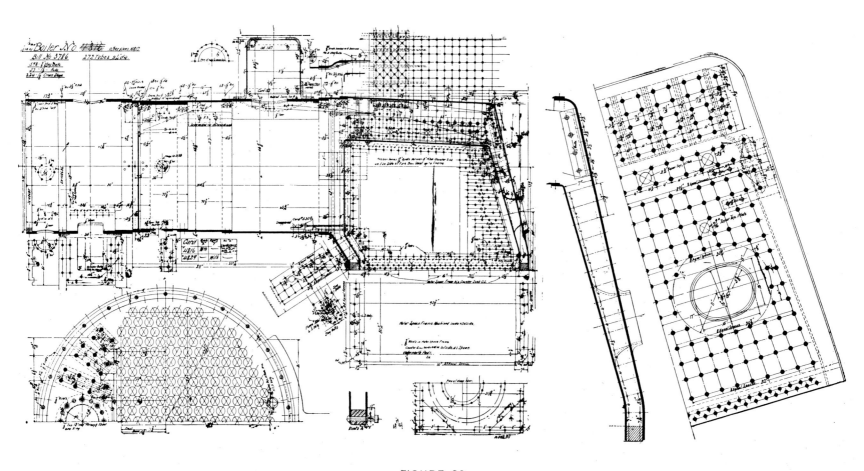

FIGURE 20

BALDWIN LOCOMOTIVE WORKS

Figure 19, shown on page 211, is a Southern Pacific ten-wheel freight locomotive. This is the first Wootten locomotive to invade the Pacific slope. It was built to burn a most inferior grade of Arizona coal, and proved a great success. This engine also proved that it could burn the high-priced coal with greater economy than the narrower firebox engines employed, and was always a free steamer.

Figure 20, shown on page 212, and Figure 21, herewith, show the Chicago, Burlington and Quincy Prairie type locomotive. At the Baldwin Locomotive Works fifty of these locomotives are now building for the Chicago, Burlington and Quincy Railroad with wide fireboxes, but as deep as possible, suitable for their peculiar fuel, which is more a lignite than a coal. If it were not encroaching on time I could describe many others.

FIGURE 21

Even New Zealand, far down in the South Pacific Ocean, demands that we shall build Wootten boilers for their three-foot six-inch gauge locomotives. We are at present building a large number for this government, capable of burning lignite, great quantities of which are found on the line of their road.

Such lines as the New York Central, Chicago and North Western, Chicago, Milwaukee and St. Paul, Buffalo, Rochester and Pittsburg, Atchison, Topeka and Sante Fe, Erie, etc., are also availing themselves of this design.

BALDWIN LOCOMOTIVE WORKS

FIGURE 22

Figure 22, illustrated herewith, Pennsylvania Railroad's No. 269, Class E-2, is one of the handsomest and best locomotives of the Atlantic type, and fitted with a wide or modified Wootten firebox. The picture is no doubt familiar to most of you, and its performance in service has been a source of great pride to those of us here to-night who delight in the achievements and progress of the Pennsylvania Railroad Company.

COMPOUND TEN-WHEEL PASSENGER LOCOMOTIVE WITH WOOTTEN BOILER FOR THE LEHIGH VALLEY RAILROAD COMPANY

DENVER & RIO GRANDE RAILROAD—THE EARLIEST AND LATEST

Locomotive Ute, the earliest, formed part of the original equipment of the road, and was built by the Baldwin Locomotive Works in 1871.
Locomotive No. 904, the latest, is one of the largest locomotives now in use on the road, and was built by the Baldwin Locomotive Works in 1901.

215

BALDWIN LOCOMOTIVE WORKS

Atlantic Type Locomotive

for the

Class 10-32¼-C-39

Gauge 4' 8½''

Buffalo, Rochester & Pittsburg Railway Company

GENERAL DIMENSIONS

CYLINDERS

Diameter	19½''
Stroke	26''
Valve	Balanced

BOILER

Diameter	64''
Thickness of Sheets	11⁄16'' and ¾''
Working Pressure	200 lbs.
Fuel	Coal

FIREBOX

Material	Steel
Length	102⅛''
Width	66⅛''
Depth, Front	72¼''
" Back	70¼''
Thickness of Sheets, Sides	5⁄16''
" " " Back	⅜''
" " " Crown	⅜''
" " " Tube	½''

TUBES

Material	Iron
Number	336
Diameter	2''
Length	15' 10''

HEATING SURFACE

Firebox	179.0 sq. ft.
Tubes	2771.0 sq. ft.
Total	2950.0 sq. ft.
Grate Area	46.9 sq. ft.

DRIVING WHEELS

Diameter Outside	72''
" of Center	66''
Journals	9'' x 12''

ENGINE TRUCK WHEELS

Diameter	33''
Journals	6'' x 10''

TRAILING WHEELS

Diameter	42''
Journals	8½'' x 12''

WHEEL BASE

Driving	7' 0''
Rigid	14' 2''
Total Engine	25' 1''
Total Engine and Tender	53' 8''

WEIGHT

On Driving Wheels	87,560 lbs.
On Truck	31,200 lbs.
" " Trailing Wheels	31,540 lbs.
Total Engine	150,300 lbs.
Total Engine and Tender	270,000 lbs.

TENDER

Diameter of Wheels	33½''
Journals	5½'' x 10''
Tank Capacity	6,000 gals.

SERVICE

Passenger.

Record of Recent Construction No. 28.   Page 7

Code Word, RECINXISTI

BALDWIN LOCOMOTIVE WORKS

Consolidation Locomotive with Vanderbilt Boiler

Class 10-38-E-338

for the

Gauge 4' 9"

Buffalo, Rochester & Pittsburg Railway Company

GENERAL DIMENSIONS

CYLINDERS

Diameter . . . . . . 22"
Stroke . . . . . 28"
Valve . . . . Balanced

BOILER (Vanderbilt)

Diameter . . . . . 66"
Thickness of Sheets . 11/16" and 3/4"
Working Pressure . . 200 lbs.
Fuel . . . . . Coal

FIREBOX

Material . . . . . Steel
Length . . . . . 94"
Width . . . . . 57"
Length of Corrugated Tube . 131"
Diameter " " . 63 3/8"
Thickness of Corrugated Sheet . 3/4"
" " Tube Sheet . 1/2"

TUBES

Material . . . . . Iron
Number 377
Diameter . . . . 2"
Length . . . . . 12' 6"

HEATING SURFACE

Firebox . . . . 135 sq. ft.
Tubes . . . . 2450 sq. ft.
Total . . . . 2585 sq. ft.
Grate Area . . . 33 sq. ft.

DRIVING WHEELS

Diameter Outside . . . 56"
"   of Center . . 50"
Journals . . . . 9" x 10"

ENGINE TRUCK WHEELS

Diameter . . . . . 30"
Journals . . . . 6" x 10"

WHEEL BASE

Driving . . . . 15' 3"
Total Engine . . . 23' 11"
Total Engine and Tender . 53' 7 3/4"

WEIGHT

On Driving Wheels . . 151,900 lbs.
On Truck . . . 17,700 lbs.
Total Engine . . . 169,600 lbs.
Total Engine and Tender . 289,000 lbs.

TENDER

Diameter of Wheels . . 33"
Journals . . . . 5 1/2" x 10"
Tank Capacity . . 6,000 gals.
"   "   (coal) . 10 tons

SERVICE

Freight.

EL PASO-ROCK ISLAND ROUTE

BALDWIN LOCOMOTIVE WORKS

Compound Consolidation Locomotive

for the

El Paso & Rock Island Railway Company

Class 10 $\frac{28}{50}$-E-85

Gauge 4' 8½"

GENERAL DIMENSIONS

CYLINDERS

Diameter (High Pressure) . .	17"
" (Low Pressure) . .	28"
Stroke	30"
Valve	Balanced Piston

BOILER

Diameter	74"
Thickness of Sheets . . .	¾"
Working Pressure . . .	200 lbs.
Fuel	Soft Coal

FIREBOX

Material	Steel
Length	120⅛"
Width	41⅞"
Depth, Front	65½"
" Back	61"
Thickness of Sheets, Sides .	⅜"
" " " Back . .	⅜"
" " " Crown .	⅜"
" " " Tube .	½"

TUBES

Material	Iron
Number	337
Diameter	2"
Length	14' 0"

HEATING SURFACE

Firebox	172.2 sq. ft.
Tubes	2453.7 sq. ft.
Total	2625.9 sq. ft.
Grate Area . . .	35.0 sq. ft.

DRIVING WHEELS

Diameter Outside . . .	60"
" of Center . .	54"
Journals . . .	9" x 11"

ENGINE TRUCK WHEELS

Diameter	30"
Journals . . .	6" x 10"

WHEEL BASE

Driving	15' 9"
Total Engine . . .	24' 4"
Total Engine and Tender .	53' 2"

WEIGHT

On Driving Wheels . .	155,555 lbs.
On Truck . . .	24,840 lbs.
Total Engine . . .	180,395 lbs.
Total Engine and Tender	300,000 lbs.

TENDER

Diameter of Wheels . . .	33"
Journals . . .	5½" x 10"
Tank Capacity . . .	6,000 gals.

SERVICE

Freight.

BALDWIN LOCOMOTIVE WORKS

Six-Coupled Double Ender Locomotive
for the

Class 12-26¼-D-14

Gauge 3' 6"

Government Railways of New Zealand

GENERAL DIMENSIONS

CYLINDERS

Diameter	16"
Stroke	22"
Valve	Balanced Piston

BOILER

Diameter	54"
Thickness of Sheets . .	9/16"
Working Pressure . . .	200 lbs.
Fuel . . Poor Lignite, analysis furnished	

FIREBOX

Material	Steel
Length	96"
Width	60"
Depth, Front . . .	53⅛"
" Back . . .	50⅛"
Thickness of Sheets, Sides .	5/16"
" " " Back .	⅜"
" " " Crown .	⅜"
" " " Tube .	½"

TUBES

Material	Iron
Number	188
Diameter	2"
Length	16' 0"

HEATING SURFACE

Firebox . . .	105.2 sq. ft.
Tubes . . .	1568.0 sq. ft.
Total . . .	1673.2 sq. ft.
Grate Area . . .	40.0 sq. ft.

DRIVING WHEELS

Diameter Outside . . .	49"
" of Center . .	44"
Journals . . .	6½" x 7"

ENGINE TRUCK WHEELS

Diameter	26"
Journals . . .	4½" x 7½"

TRAILING WHEELS

Diameter	30"
Journals . . .	5½" x 9½"

WHEEL BASE

Driving	9' 2"
Total Engine . . .	26' 3"
Total Engine and Tender .	48' 3"

WEIGHT

On Driving Wheels . .	64,530 lbs.
On Truck . . .	16,800 lbs.
" " Trailing Wheels .	17,400 lbs.
Total Engine . . .	98,730 lbs.
Total Engine and Tender	149,000 lbs.

TENDER

Diameter of Wheels . .	28"
Journals . . .	3¾" x 7"
Tank Capacity . .	2,000 gals.

SERVICE

Passenger.
Curves, 330 feet radius.

Record of Recent Construction No. 28.   Page 13

Code Word, RECIPERO

BALDWIN LOCOMOTIVE WORKS

Consolidation Locomotive

Class 10-36-E-414

for the

Gauge 4' 8½''

Choctaw, Oklahoma & Gulf Railroad Company

GENERAL DIMENSIONS

CYLINDERS

Diameter	21''
Stroke	26''
Valve	Balanced

BOILER

Diameter	64''
Thickness of Sheets	9/16'' and 5/8''
Working Pressure	160 lbs.
Fuel	Soft Coal

FIREBOX

Material	Steel
Length	103 3/16''
Width	33 3/8''
Depth, Front	67 3/4''
" Back	65 3/4''
Thickness of Sheets, Sides	5/16''
" " " Back	5/16''
" " " Crown	3/8''
" " " Tube	1/2''

TUBES

Material	Iron
Number	241
Diameter	2 1/4''
Length	13' 9 3/4''

HEATING SURFACE

Firebox	160.8 sq. ft.
Tubes	1948.4 sq. ft.
Total	2109.2 sq. ft.
Grate Area	23.84 sq. ft.

DRIVING WHEELS

Diameter Outside	56''
" of Center	50''
Journals	8'' x 8 1/2''

ENGINE TRUCK WHEELS

Diameter	29 1/4''
Journals	5'' x 8''

WHEEL BASE

Driving	15' 0''
Total Engine	23' 1''
Total Engine and Tender	50' 7 1/2''

WEIGHT

On Driving Wheels	124,165 lbs.
On Truck	17,780 lbs.
Total Engine	141,945 lbs.
Total Engine and Tender	222,000 lbs.

TENDER

Diameter of Wheels	33''
Journals	4 1/4'' x 8''
Tank Capacity	4,000 gals.

SERVICE

Freight.

BALDWIN LOCOMOTIVE WORKS

Ten-Wheel Locomotive

for the

Southern Pacific Company

Class 10-32-D-681

Gauge 4' 8½"

GENERAL DIMENSIONS

CYLINDERS

Diameter	19"
Stroke	26"
Valve	Balanced

BOILER

Diameter	62"
Thickness of Sheets	11/16"
Working Pressure	200 lbs.
Fuel, Sonora Anthracite Coal, as sample furnished by company.	

FIREBOX

Material	Steel
Length	114"
Width	96"
Depth, Front	49½"
" Back	47½"
Thickness of Sheets, Sides	5/16"
" " " Back	5/16"
" " " Crown	3/8"
" " " Tube	½"

TUBES

Material	Iron
Number	224
Diameter	2"
Length	14' 0"

HEATING SURFACE

Firebox	176 sq. ft.
Tubes	1627 sq. ft.
Total	1803 sq. ft.
Grate Area	76 sq. ft.

DRIVING WHEELS

Diameter Outside	63"
" of Center	56"
Journals	8" x 10"

ENGINE TRUCK WHEELS

Diameter	30"
Journals	5" x 10"

WHEEL BASE

Driving	12' 6"
Total Engine	23' 4"
Total Engine and Tender	49' 8"

WEIGHT

On Driving Wheels	117,330 lbs.
On Truck	33,200 lbs.
Total Engine	150,530 lbs.
Total Engine and Tender	230,000 lbs.

TENDER

Diameter of Wheels	33"
Journals	4¼" x 8"
Tank Capacity	4,000 gals.

SERVICE

Passenger and Freight.

BALDWIN LOCOMOTIVE WORKS

Compound Ten-Wheel Locomotive

Class 10 $\frac{25}{46}$-D-106

for

Gauge 4' 8½"

Chicago & Alton Railway Company

GENERAL DIMENSIONS

CYLINDERS

Diameter (High Pressure) .	15½"
" (Low Pressure) . .	26"
Stroke	28"
Valve . . . . Balanced Piston	

BOILER

Diameter	68"
Thickness of Sheets . 1¹¹⁄₁₆" and ¾"	
Working Pressure . .	200 lbs.
Fuel	Soft Coal

FIREBOX

Material	Steel
Length	120⅛"
Width	41⅞"
Depth, Front . .	76¾"
" Back . . .	74¾"
Thickness of Sheets, Sides .	⁵⁄₁₆"
" " " Back .	⅜"
" " " Crown .	⅜"
" " " Tube .	½"

TUBES

Material	Iron
Number	408
Diameter	2"
Length	15' 6"

HEATING SURFACE

Firebox	187.0 sq. ft.
Tubes	3293.5 sq. ft.
Total	3480.5 sq. ft.
Grate Area . . .	34.9 sq. ft.

DRIVING WHEELS

Diameter Outside . . .	68"
" of Center . .	62"
Journals . . .	9½" x 13"

ENGINE TRUCK WHEELS

Diameter	32¼"
Journals . . .	6½" x 13"

WHEEL BASE

Driving	14' 5"
Rigid	7' 6"
Total Engine . . .	26' 7"
Total Engine and Tender .	55' 3½"

WEIGHT

On Driving Wheels .	131,225 lbs.
On Truck . . .	49,530 lbs.
Total Engine . .	180,755 lbs.
Total Engine and Tender .	300,000 lbs.

TENDER

Diameter of Wheels . .	35¼"
Journals	5" x 9"
Tank Capacity . .	6,000 gals.

SERVICE

Passenger.

BALDWIN LOCOMOTIVE WORKS

Atlantic Type Locomotive

for the

Gauge 4' 9"

Paris, Lyons & Mediterranean Railway

GENERAL DIMENSIONS

CYLINDERS

Diameter	17¼"
Stroke	26"
Valve	Balanced Piston

BOILER

Diameter	58"
Thickness of Sheets . . .	11/16"
Working Pressure . .	213 lbs.
Fuel	Coal

FIREBOX

Material	Copper
Length	120"
Width	42"
Depth, Front	71¼"
" Back . . .	67½"
Thickness of Sheets, Sides . .	5/8"
" " " Back .	5/8"
" " " Crown . .	5/8"
" " " Tube	7/8" and 5/8"

TUBES

Material	Iron
Number	246
Diameter	2"
Length	15' 1"

HEATING SURFACE

Firebox	170.4 sq. ft.
Tubes	1925.4 sq. ft.
Total	2095.8 sq. ft.
Grate Area . . .	35.0 sq. ft.

DRIVING WHEELS

Diameter Outside . . .	84¼"
" of Center . .	78"
Journals	8" x 10"

ENGINE TRUCK WHEELS

Diameter	36"
Journals	5½" x 10"

TRAILING WHEELS

Diameter	54¼"
Journals	8" x 10"

WHEEL BASE

Driving	7' 3"
Rigid	14' 6"
Total Engine	26' 8"
Total Engine and Tender .	52' 0"

WEIGHT

On Driving Wheels . .	70,970 lbs.
On Truck . . .	33,790 lbs.
" " Trailing Wheels	31,400 lbs.
Total Engine . .	136,160 lbs.
Total Engine and Tender	226,000 lbs.

TENDER

Diameter of Wheels . .	47¼"
Journals	5" x 9½"
Tank Capacity . .	4,490 gals.

SERVICE

Express Passenger.

BALDWIN LOCOMOTIVE WORKS

Consolidation Locomotive

for the

Western Maryland Railroad Company

Class 10-38-E-311

Gauge 4' 8½"

GENERAL DIMENSIONS

CYLINDERS

Diameter	22"
Stroke	28"
Valve	Balanced

BOILER

Diameter	68"
Thickness of Sheets . . .	11/16"
Working Pressure . . .	180 lbs.
Fuel	Soft Coal

FIREBOX

Material	Steel
Length	118"
Width	41"
Depth, Front	71"
" Back	67"
Thickness of Sheets, Sides .	3/8"
" " " Back .	3/8"
" " " Crown .	1/2"
" " " Tube .	1/2"

TUBES

Material	Iron
Number	247
Diameter	2¼"
Length	14' 6"

HEATING SURFACE

Firebox	191.0 sq. ft.
Tubes	2094.0 sq. ft.
Total	2285.0 sq. ft.
Grate Area	33.6 sq. ft.

DRIVING WHEELS

Diameter Outside . . .	56"
" of Center . .	50"
Journals . . .	8½" x 10"

ENGINE TRUCK WHEELS

Diameter	30"
Journals . . .	5" x 10"

WHEEL BASE

Driving	15' 4"
Total Engine . . .	23' 8"
Total Engine and Tender .	52' 8"

WEIGHT

On Driving Wheels . .	141,682 lbs.
On Truck . . .	17,277 lbs.
Total Engine . . .	158,959 lbs.
Total Engine and Tender	278,000 lbs.

TENDER

Diameter of Wheels . .	33"
Journals	5" x 9"
Tank Capacity . .	6,000 gals.

SERVICE

Freight.

Code Word, RECIPITORE

BALDWIN LOCOMOTIVE WORKS

Mogul Locomotive

Class 8-22-D-264

for the

Gauge 4' 8½"

Catskill & Tannersville Railroad Company

GENERAL DIMENSIONS

CYLINDERS

Diameter	14″
Stroke	20″
Valve	Balanced

BOILER

Diameter	48″
Thickness of Sheets	½″
Working Pressure	180 lbs.
Fuel	Hard Coal.

FIREBOX

Material	Steel
Length	42 ³⁄₁₆″
Width	50⅝″
Depth, Front	54″
" Back	38½″
Thickness of Sheets, Sides	⁵⁄₁₆″
" " " Back	⁵⁄₁₆″
" " " Crown	⅜″
" " " Tube	½″

TUBES

Material	Iron
Number	III
Diameter	2″
Length	13' 4″

HEATING SURFACE

Firebox	60.5 sq. ft.
Tubes	770.0 sq. ft.
Total	830.5 sq. ft.
Grate Area	15.4 sq. ft.

DRIVING WHEELS

Diameter Outside	42″
" of Center	36″
Journals	6″ x 7″

ENGINE TRUCK WHEELS

Diameter	24″
Journals	4″ x 6¼″

WHEEL BASE

Driving	8' 6″
Total Engine	15' 3″
Total Engine and Tender	38' 9½″

WEIGHT

On Driving Wheels	59,910 lbs.
On Truck	7,250 lbs.
Total Engine	67,160 lbs.
Total Engine and Tender	100,000 lbs.

TENDER

Diameter of Wheels	24″
Journals	3¼″ x 5″
Tank Capacity	1,600 gals.

SERVICE

Passenger and Freight.

Length of road, 6 miles. Grades, five per cent. reverse without intervening level. Sharpest curve, 30 degrees.

BALDWIN LOCOMOTIVE WORKS

Four-Coupled Locomotive

for the

Berlin Mills Company

Class 4-14-C-129

Gauge 4' 8½"

GENERAL DIMENSIONS

CYLINDERS

Diameter	10"
Stroke	14"
Valve	Balanced

BOILER

Diameter	32"
Thickness of Sheets . . .	⅜"
Working Pressure . .	160 lbs.
Fuel	Soft Coal

FIREBOX

Material	Steel
Length	36⁷⁄₁₆"
Width	36⅝"
Depth	40¾"
Thickness of Sheets, Sides .	⁵⁄₁₆"
" " " Back .	⁵⁄₁₆"
" " " Crown .	⁵⁄₁₆"
" " " Tube .	½"

TUBES

Material	Iron
Number	87
Diameter . . .	1½"
Length . . .	8' 4½"

HEATING SURFACE

Firebox . . .	39.5 sq. ft.
Tubes . . .	281.6 sq. ft.
Total . . .	321.1 sq. ft.
Grate Area . . .	9.3 sq. ft.

DRIVING WHEELS

Diameter Outside . . .	34"
" of Center . .	26"
Journals . . .	5" x 6¼"

WHEEL BASE

Driving	5' 0"
Total . . .	5' 0"

WEIGHT

On Driving Wheels . .	34,070 lbs.
Total . . .	34,070 lbs.

TANK

Tank Capacity . . .	450 gals.

SERVICE

Switching

Curves, 70 feet radius.

BALDWIN LOCOMOTIVE WORKS

Four-Coupled Locomotive

Class 4-26-C-80

for the

Gauge 4' 8½"

Clarendon & Pittsford Railroad Company

GENERAL DIMENSIONS

CYLINDERS

Diameter	16"
Stroke	22"
Valve	Balanced

BOILER

Diameter	54"
Thickness of Sheets . .	½"
Working Pressure . .	160 lbs.
Fuel	Soft Coal

FIREBOX

Material	Steel
Length	57 3/16"
Width	42"
Depth, Front	52½"
" Back . . .	49¾"
Thickness of Sheets, Sides .	5/16"
" " " Back .	5/16"
" " " Crown .	3/8"
" " " Tube .	½"

TUBES

Material	Iron
Number	190
Diameter	2"
Length	11' 6"

HEATING SURFACE

Firebox . . .	61.2 sq. ft.
Tubes . . .	1135.7 sq. ft.
Total . . .	1196.9 sq. ft.
Grate Area . . .	16.6 sq. ft.

DRIVING WHEELS

Diameter Outside . . .	46"
" of Center . .	40"
Journals	8" x 8½"

WHEEL BASE

Driving	8' 0"
Total Engine . . .	8' 0"
Total Engine and Tender .	25' 9"

WEIGHT

On Driving Wheels . .	75,405 lbs.
Total Engine . . .	75,405 lbs.
Total Engine and Tender .	115,000 lbs.

TENDER

Diameter of Wheels . .	30"
Journals	4¼" x 8"
Tank Capacity . .	2,000 gals.

SERVICE

Switching.

BALDWIN LOCOMOTIVE WORKS

Four-Coupled Locomotive

Class 4-18-C-59

for the

Gauge 3' 0"

Colorado Fuel and Iron Company

GENERAL DIMENSIONS

CYLINDERS

Diameter	12"
Stroke	16"
Valve	Plain D

BOILER

Diameter	40"
Thickness of Sheets	7/16"
Working Pressure	160 lbs.
Fuel	Soft Coal

FIREBOX

Material	Steel
Length	36 7/16"
Width	35 1/4"
Depth	45 1/2"
Thickness of Sheets, Sides	3/8"
" " " Back	5/16"
" " " Crown	3/8"
" " " Tube	1/2"

TUBES

Material	Iron
Number	127
Diameter	1 1/2"
Length	8' 1 5/8"

HEATING SURFACE

Firebox	45.0 sq. ft.
Tubes	401.0 sq. ft.
Total	446.0 sq. ft.
Grate Area	8.9 sq. ft.

DRIVING WHEELS

Diameter Outside	33"
" of Center	28"
Journals	5 1/2" x 6"

WHEEL BASE

Driving	5' 0"
Total	5' 0"

WEIGHT

On Driving Wheels	45,160 lbs.
Total	45,160 lbs.

TANK

Tank Capacity	500 gals.

SERVICE

Furnace Switching.

241

BALDWIN LOCOMOTIVE WORKS

Mogul Locomotive

Class 8-24-D-118

for the

Gauge 3' 0"

United Collieries Company for Altoona Division P. J. E. & E. R. R.

GENERAL DIMENSIONS

CYLINDERS

Diameter	15"
Stroke	20"
Valve	Balanced

BOILER

Diameter	54"
Thickness of Sheets	½"
Working Pressure	160 lbs.
Fuel	Soft Coal

FIREBOX

Material	Steel
Length	83³⁄₁₆"
Width	24"
Depth, Front	56¾"
" Back	46"
Thickness of Sheets, Sides	⁵⁄₁₆"
" " " Back	⁵⁄₁₆"
" " " Crown	⅜"
" " " Tube	½"

TUBES

Material	Iron
Number	210
Diameter	1¾"
Length	9' 2"

HEATING SURFACE

Firebox	98.4 sq. ft.
Tubes	873.8 sq. ft.
Total	972.2 sq. ft.
Grate Area	13.84 sq. ft.

DRIVING WHEELS

Diameter Outside	40"
" of Center	34"
Journals	6½" x 7"

ENGINE TRUCK WHEELS

Diameter	26"
Journals	4" x 6"

WHEEL BASE

Driving	10' 6"
Total Engine	17' 6"
Total Engine and Tender	38' 7½"

WEIGHT

On Driving Wheels	62,600 lbs.
On Truck	9,850 lbs.
Total Engine	72,450 lbs.
Total Engine and Tender	112,000 lbs.

TENDER

Diameter of Wheels	26"
Journals	3¼" x 6"
Tank Capacity	2,000 gals.

SERVICE

Freight.

Guaranteed to haul 100 tons (of 2,000 lbs.) up a straight grade of four per cent. Track and cars being in good condition.

AT NIGHT

—

The Baldwin
Locomotive
Works
as seen from
Broad and
Spring Garden
Streets

BALDWIN LOCOMOTIVE WORKS

The Building of a Modern Locomotive

From the Brotherhood of Locomotive Firemen's Magazine

February, 1901

Construction of Boiler

IN order to properly understand the building of a modern locomotive, and comprehend the many operations necessary to produce the finished machine, it is essential to visit a large locomotive works, and actually view the process as there carried on. This article will only attempt, in a general way, to outline the process with the help of a number of illustrations which, it is hoped, will make the description clearer.

The pictures here shown are from photographs taken at the Baldwin Locomotive Works, in Philadelphia, and the accompanying description applies to the methods employed at those shops. It must suffice, in this con-

nection, to note that these works are the largest of their kind in the world; that they employ over 8000 men; have a rated capacity of 1000 locomotives per annum, and actually turned out over 1200 in the year just closed, at the end of which the total output had been about 18,500 engines, over 1500 of them being four-cylinder compounds. The history of the works, since their founding by Matthias Baldwin in 1831, has been one of real progress from the start, and to-day Baldwin locomotives are doing effective work all over the world. Lack of space, unfortunately, prevents our taking up, in detail, the subject of shop practice as employed at these works.

BALDWIN LOCOMOTIVE WORKS

In describing the building of a locomotive, we will take up the subject under four heads, viz.:— First, boilers; second, cylinders; third, frames, wheels and other parts, and fourth, assembling or erecting.

While all parts of the locomotive have undergone a

remarkable development during recent years, it is the boiler which, to-day, is receiving the greatest amount of attention. It is becoming more and more recognized that the ability of a locomotive to maintain a high average speed on long hauls, depends chiefly upon the capacity of its boiler; and the problem of getting increased boiler power—and especially more grate area—is one of the knotty questions confronting the up-to-date locomotive designer.

Steel plate is used exclusively in modern American locomotive boilers, although in many engines for foreign export the firebox is of copper. The steel is received at the works in sheets of various sizes and thicknesses, some of them over twenty feet long, this size being required to form a ring for the very large boilers now in use. The plates are moved about in the shop by means of overhead traveling cranes; this arrangement being made easier since most of the tools are driven by separate electric motors, thus avoiding overhead belts.

The first operation is to prepare the plates for punching and drilling. The sheet is laid on a table, and the centers of the holes marked out by means of standard

FIG. 1. MULTIPLE DRILLING MACHINE

gauges. In cases where a plate is to be flanged, the holes for the flange are not marked off until after that operation is completed, otherwise a perfect match with corresponding holes in other plates could not be assured. Straight lines, along which rivet holes are to lie, are first laid down, the centers of the holes being then carefully spaced, so that their position can be determined with absolute accuracy. In the case of sheets for the barrel of the boiler, plates which lap are divided into the same number of equal spaces, and the rivet holes located at the points of division, so that, provided the lengths of the sheets are correctly laid off, the holes are sure to match.

The centers having been accurately located, the next operation is to form the holes, which is done either by punching or drilling, according to specifications. Foreign orders require that all holes must be drilled; but in American practice it is customary to use the punch to a large extent. In all cases where the punch is used, the holes are made smaller than the required diameter, and subsequently reamed out to size. All flanged plates are annealed, and all flue sheets after punching are straightened and annealed, and then the holes are reamed

No. 2. HYDRAULIC FLANGING PRESS

to gauges by special machinery. In the case of rivet and stay-bolt holes, the reaming is done with small portable pneumatic machines.

Many of the punching machines are driven by separate electric motors, and they are of various designs

BALDWIN LOCOMOTIVE WORKS

to handle different sizes of plates. Above each machine is a crane, from which the plate is hung by means of chains. The punch has a projection on its under side; several men guide the plate until this projection catches in one of the center punch marks, when, by moving a lever, the machine is put in operation, and promptly bites out the metal.

Figure 1 shows a large multiple drilling machine, which is boring through five sheets, clamped together so that the holes will match. Where a piece is to be cut out for a dome, for example, the plate is drilled around the edge, and the metal so removed. A hole has been so cut in the sheet seen in the left-hand corner of the illustration.

All sheets on which flanging is to be done are now sent to the flanging shop. As far as possible, flanges are made on hydraulic presses, one of which is shown in Figure 2. This machine is operated by means of two large "accumulators," and can exert a maximum pressure of 365 tons. The plate is heated in a large furnace, part of which is seen on the right-hand side of the illustration. It is then placed on a suitable form,

clamped to the lower table, a corresponding form having been clamped to the under side of the upper table, as shown in the picture. The lower table is then raised by hydraulic power, and the entire flange made at one heat. Tube sheets, furnace-door flanges, smokebox fronts, dome rings, and many other parts can be formed on this machine with a rapidity and an accuracy impossible in hand work. In working a back head, the flange around the edge of the plate, and that around the furnace door, can both be formed at one heat.

Flanges of odd shape, for which there are no dies to form them on the press, are made by hand. The edge of the plate is placed in a fire forced by an air blast, and, when the desired heat is obtained, is laid on a form and bent by means of heavy wooden mallets. In bending on a straight edge, the plate is clamped to the form, and one heat usually suffices for the operation. Flanging by hand is naturally a slow process, as three feet is about the greatest length that can be heated and bent at once; and it is only when suitable dies for the hydraulic press cannot be profitably made, that this system is employed.

BALDWIN LOCOMOTIVE WORKS

The plates are now sent back to the boiler shop where they are prepared for assembling. The roughened edges of the flanges, whenever it is possible to do so, are planed off on special machines; but in most cases the edges are dressed by chipping with a hammer and chisel. In the meantime, the plates for the barrel of the boiler have been trimmed off in a large sheering press, and the edges then planed on a plate planer, after which they are ready for the bending rolls. Each of these machines consists of three rolls, power to turn them being furnished by an electric motor. The two lower ones are on the same plane, while the upper one is adjustable as to height, its position being controlled by large screws operated by gears, one screw being placed at each end. By altering the position of the upper roll, the plate can be bent to any desired radius. The workmen use a sheet-iron templet as a gauge, and can tell accurately when the proper curvature has been reached.

The plates, having been bent and flanged, are ready to be assembled for the riveting machines. The rivet holes in the flanges are laid off by temporarily assembling the piece with others which have already been punched and drilled, the holes being thus made to match exactly. Special punches and drills, with the tools traveling horizontally instead of vertically, are used to pierce the flanges.

The boiler is now assembled in two principal parts, viz.—the smokebox and two front rings, and the third ring and outer shell of the firebox. The sheets are temporarily united with bolts. The mud ring, which has been formed of wrought iron, with joints welded throughout, and drilled for riveting to the firebox, is fitted to the inside shell; which is not united with the outside shell until after the machine riveting has been finished.

The riveters used at the Baldwin works are operated with hydraulic power. Each die is placed on the top of an upright; and these, in the largest machines, are tall enough to lower the assembled boiler between them, so that the riveting can be started at one end, and worked all the way to the other by simply raising the shell, which is suspended, by means of chains, from overhead travelers. The movable die of the riveter is always on the outside of the boiler shell, while the rivet head is formed against the fixed die on the inside. These

BALDWIN LOCOMOTIVE WORKS

FIG. 3. BOILER RIVETING MACHINE

The shell is hung, by means of chains, from travelers placed under the roof of the shop, and its manipulation is a very simple matter, as it is perfectly free to be raised or lowered, or turned to any desired position. The illustration shows plainly the cylinder and piping for conveying water to the ram of the movable die; while behind the operator is the furnace in which the rivets are heated.

Figure 4 shows a group of shells and plates ready for the riveting machines, several of which are seen in the illustration. Those in the middle ground are the tallest. The machine on the left has been working on a boiler for a Pennsylvania Railroad Consolidation locomotive, the barrel of which has been riveted to the outside shell of the firebox. This boiler is of the Belpaire type, and the back head for it is seen standing beside the small riveter in the lower right-hand corner of the illustration. The rivet holes in the flange have already been punched. On the right is seen a boiler with a wide, or modified Wootten, firebox, for a Baltimore and Ohio compound freight engine. The barrel has not yet been riveted to the firebox, the two being united, as shown in the

machines are practically noiseless. Each machine has its own furnace for heating the rivets. Three men swing the boiler into position and handle the riveter, while a boy looks after the furnace. One of these machines, with a boiler standing beside it, is shown in Figure 3.

illustration, by means of bolts and nuts. The building of this type of firebox has been made more or less of a specialty at the Baldwin works. It is possible to put ninety square feet of grate area into a boiler of this pattern, and, with such a large firebox, steam can be raised successfully with very inferior grades of fuel. Other views of this boiler will be shown in the paper on "Erecting."

When the boiler leaves the machine, most of the riveting has been done, although the inside shell of the firebox has not been united with the outside, and the front and back sections are still separated.

A twenty-five-ton traveling crane now hoists the several parts to the second floor of the shop, a view of which is shown in Figure 5. The rear half of the boiler is placed upside down, and the inside shell of the firebox, with the mud ring and the back head, are assembled into it by means of bolts and nuts. The mud ring is riveted by means of a small hydraulic machine, which is hung from a crane so that it can be easily moved about. One of these riveters can be plainly seen, suspended above a boiler on the left-hand side of the illustration.

In cases where the back head is flanged *outward*, as is often done in wide firebox boilers, the machine riveter can be used in securing it to the shell; but where it is flanged *inward*, hand riveting must be resorted to, as the inner ends of the rivets are not accessible to the machine.

FIG. 4. BOILER SHELLS AND RIVETING MACHINES

The inside firebox, mud ring and back head having been riveted to the outside shell, the boiler is placed on suitable pedestals, right side up, and the front half of the barrel—previously finished—is assembled with it. This operation is sometimes performed while the boiler is still on its back, as in the case of those shown in the foreground of Figure 5. These boilers, it may be noted, have only two rings in the barrel, and in the boiler on the extreme left they have not yet been riveted together. The seam where the halves unite is riveted up by hand, the heads being formed on the outside, in a die which is struck with heavy hammers. In the meantime, the stay-bolt holes are being threaded with small portable pneumatic tapping machines. The stay bolts are of iron, threaded at each end, the middle being turned down to a diameter equal to the inside diameter of the thread. In the center of the bolt is a small hole, leakage through which at once occurs if the bolt breaks. On the outer end of each bolt is a square head, with which it is screwed in by means of a special wrench. The head is then cut off with a special machine, and the bolt hammered down at each end, while cold, with blows from a heavy hammer. In foreign work, where copper fireboxes are frequently used, the stay bolts are usually of copper, but similar, in all other respects, to the iron ones just described.

FIG. 5. BOILER SHOP

BALDWIN LOCOMOTIVE WORKS

The firebox in a modern boiler is usually radial-stayed all around, although, in some types, the crown sheet is still supported by means of crown bars and sling stays, the saddles for the stays having been machine riveted to the roof and bolted to the crown sheets, previous to assembling.

The dome, in the meantime, has been riveted to a flanged ring of pressed steel, previously united with the shell by the hydraulic machine. The body of the dome is formed of a sheet of boiler plate, which has been punched for riveting to the dome top, then bent to the proper radius, and lap-welded on a form by blows struck with heavy hammers. The dome top and base ring are of pressed steel, being formed on a small hydraulic machine. The cap is then drilled to match the holes in the body of the dome, after which the two are assembled, the holes being reamed out together with a pneumatic drill, and subsequently riveted on a hydraulic riveter. The dome base is accurately machined to fit the boiler and is drilled in position.

As the plates are assembled, the sheets are calked at the joints to prevent leakage. This is done by

FIG. 6. BOILER IN ERECTING SHOP

means of a tool with a rounded edge, which forms a fillet and pinches the plates tightly together without cutting the metal. This method of calking, known as the "concave," was devised by Mr. J. W. Connery, of the Baldwin works, and has proved highly successful.

BALDWIN LOCOMOTIVE WORKS

The work on the boiler shell is now practically completed. The riveting is finished; the stay bolts are all screwed in; the dome has been mounted, and the tube sheets inserted. The boiler is now sent to the erecting shop, Figure 6 showing its appearance as it is received there. The boiler is intended for a large compound Consolidation engine. The ring just in front of the dome is hand riveted; practically all the rest of the work has been done by machine, except that around the back head, which sheet, as the illustration shows, is flanged inward, thus preventing the use of the machine riveter. It will be noticed that nearly all the seams, including the mud ring, are double riveted. The arrangement of a butt joint, with four rows of rivets through the outside covering strip, and six through the inside, is plainly shown just back of the smokebox. The extension front is secured to the barrel by an internal ring, so that there is a smooth finish on the outside. It will be noticed that this boiler, as is usually the case in modern practice, is built "telescopic," each ring in the barrel lapping on the inside of the one next back of it, the smokebox ring excepted.

It yet remains to insert the tubes and various mountings, and to test the boiler for leakage. These operations will be described in the paper on "Erecting."

The next section of the article will deal with the locomotive cylinder and its method of construction.

BALDWIN LOCOMOTIVE WORKS

The Cylinders

ALTHOUGH pressed and cast steel are now being very extensively used in locomotive construction, there is one part which is universally made of cast iron, and that is the cylinders.

The foundry at the Baldwin works, where cylinders, wheel centers and other parts are cast, extends the length of a city block, and is shown in Figure 7. Down the center of the shop are placed seven jib cranes, and in addition there are several small traveling cranes, to assist in handling the molds and ladles.

There are three cupolas, each having a capacity of fifty tons of iron at one heat; the amount actually used in this shop being about 135 tons per day.

Cast iron is graded by chemical analysis only, and that used for cylinders contains only about one and seven tenths per cent. of silicon, as it is necessary to have this part of as hard a material as can conveniently be worked. The grade of iron is regulated by the relative amounts of "pig iron" and "scrap" which are mixed to produce it. Before using the pig iron, samples are sent to the laboratory for analysis, in order that its composition may be accurately determined. The scrap comes to the foundry in the shape of old and spoiled castings, which are broken into small pieces by raising a heavy weight and allowing it to drop on them. The quality of the scrap can frequently be determined with sufficient accuracy, by a simple inspection, but when this cannot be done, it is also sent to the laboratory for analysis.

One of the cupolas, mentioned above, is used exclusively to produce the iron of which cylinders are made. The cupola is simply a stack built of metal plates and lined with fire brick, and fed through a charging door located about twenty feet above the ground; the charging being done from a gallery, to which trucks loaded with pig, scrap, and coke, are hoisted by means of a crane and an elevator.

BALDWIN LOCOMOTIVE WORKS

The operation is started late in the morning by building a coke fire in the cupola, 1800 pounds of fuel being used to form a bed, on which are placed alter-nate layers of a mixture of pig and scrap, and coke. The relative amounts of pig and scrap used depend upon the result of the analysis of the same, and different proportions must be used according to their composi-tion, to produce the grade of iron desired. The coke is of the best quality, the usual ratio, by weight, of coke to iron, being about as one to eight. A small amount of limestone is added from time to time, in order to form a "flux" to carry off the impurities in the iron.

About two hours after the fire has been lighted, the cupola is filled to the bottom level of the charging door. An air blast, produced by a large blower driven by a steam engine, is now forced into the cupola through two openings, called *tuyeres*, located about five feet above the bottom, and in the course of perhaps half an hour, the cupola is ready for tapping. The tap hole is a small opening closed with a clay plug, and discharging into a spout from which the metal runs into the ladles; the slag, containing the impurities, running out through another opening, located above the tap hole and on the opposite side of the cupola.

FIG. 7. WHERE BALDWIN CYLINDERS ARE CAST
The cylinder foundry at the Baldwin Locomotive Works, one block in length.

BALDWIN LOCOMOTIVE WORKS

The molders in the meantime have been getting ready the molds in which the cylinders are cast. The high and low pressure cylinders for one of the new Baltimore and Ohio Consolidation engines are respectively fifteen and one half and twenty-six inches in diameter, and a casting for cylinders of this size and type requires a pattern built up in three sections, the lines of division running through the center lines of the high pressure cylinder and steam chest, and the center line of the low pressure cylinder, respectively. The flasks are built of cast-iron plates, bolted together at the corners, and having a trunion at each end, so that they can be conveniently handled by the crane. A number of molds of this kind, in course of erection, are shown in Figure 8. It should be explained, in this connection, that in the Vauclain compound locomotive the high and low pressure cylinders, and the steam chest, are cast together with half the saddle; the valve being of the piston type, and working in a cylindrical bushing, which is forced into the steam chest after the casting is finished, as will be explained hereafter.

The casting weighs about 8,700 pounds, and the

FIG. 8. FLASKS FOR CASTING CYLINDERS

pouring is done from a ladle having a capacity of 12,000 pounds. This ladle is not filled directly from the cupola; a smaller one is put under the tap hole, and in it the molten iron is transferred to the large ladle, two fillings of the small one being required. The slight cooling of the metal, while the small ladle is being filled a second time, is rather an advantage, as the metal is too hot to be poured as it runs from the cupola. The ladle is swung to the mold on one of the large cranes, and tilted about a horizontal axis by means of a handle connected with worm wheels. The metal is poured steadily until the mould is completely filled; then as it shrinks when it begins to cool, a little more is added to keep the mold full. The pouring being finished, the mold is allowed to stand about twelve hours, before the casting is taken out of it; and after it is removed it stands for a day before the cleaners begin their work. A great quantity of sand adheres to the casting; this is all carefully removed and projections on the surface are chipped off, giving the piece a fairly smooth and clean appearance.

The cylinder now leaves the foundry, and the work of machining it down to size is begun. The first operation is to plane off the end surfaces, so that the casting can be more accurately laid out before it is sent to the boring machine. The piece is set on a planer with the axis of the cylinder vertical, and a cut is taken off the end surfaces surrounding the cylinders and steam chest at both ends of the casting. It is then placed on a table, and the cylinders and steam chest carefully centered by means of gauges, to make sure that the casting will finish to the drawing. The cylinders, when finished, are slightly counterbored at the ends; and circles are scribed on the planed end surfaces to show the diameter of the counterbore. The diameter of the steam chest, when finished, is indicated by similar circles; after which the casting is ready to go to one of the boring machines.

These are most interesting machines, designed for boring out both cylinders and the steam chest at the same time. Figures 9 and 10 clearly show their principal features. It is seen that there are three boring bars, that for the high-pressure cylinder being fixed, while the one for the low pressure is adjustable horizontally, and for the steam chest both vertically and horizontally; so that the machine can be adjusted to fit cylinders of different sizes. The boring bars can be rotated

FIG. 9. THE BORING MACHINE

Showing operation of simultaneously boring the high-pressure cylinder, the low-pressure cylinder, and the steam chest
of a Baldwin Compound Locomotive.

BALDWIN LOCOMOTIVE WORKS

FIG. 10.   THE BORING MACHINE

Another illustration showing the operation of boring the cylinders of a Baldwin pound Locomotive.   This machine bores at the same time the low-pressure cylinder, the high-pressure cylinder, and the steam chest.

length, and in the center of the bar is placed a screw which rotates independently by means of separate gearing. The tools are carried on collars, which slip over the bars and have pieces projecting inward through the slots, and taking hold of the screws, so that as the latter rotate, the tools are "fed" along the bars. The slots and collars on two of the boring bars are plainly shown in Figure 9. It will be noticed that the collar on the low-pressure bar is arranged to carry four tools, and by placing them staggered around the collar, four separate cuts are taken at each rotation.

Before placing the cylinder on the machine, the bars are so adjusted that their centers are exactly the same distances apart as are the centers of the cylinders and steam chest to be bored. The bed on which the cylinder is to be placed is now slid out to the right (looking at the machine as in Figure 9) leaving the boring bars unsupported at their outer ends. The cylinder is placed on the bed, and conveniently supported by means of metal blocks beneath it, a jack being set under the saddle, as shown in the illustration. The bed is now slid back and the position of the cast-

independently of one another, by means of toothed clutches. Each bar has a slot running almost its entire

BALDWIN LOCOMOTIVE WORKS

ing adjusted, until the distance from the surface of the boring bar to the circles previously scribed on the end faces of the cylinder is exactly the same all the way around the circumference. This distance is measured alike at both ends of the casting, and the latter is clamped in position.

As they come from the foundry, the cylinders are about three-fourths inch less in diameter than they are to be when bored; measuring in this case, fourteen and three-fourths and twenty-five and one-fourth inches for the high and low pressure, respectively. Three cuts are now taken through the cylinders and steam chest; although in the case of the latter, four are sometimes required. In taking the first cut, three tools are usually used on the low-pressure cylinder, and two on the high pressure and steam chest. The second cut, using one tool only, brings the bore almost to finished size; the finishing cut with a broad tool being little more than a scrape. The steam chest in this style of cylinder, is formed with eight circular bridges; and if their diameters are not uniform in the rough casting, four cuts may be necessary. Great care must be taken to obtain

exactly the right diameter; which is measured by means of calipers accurately set to gauges. Before the casting is removed from the machine the end faces of the cylinders are turned up, and the counterboring finished.

A considerable amount of work remains to be done on the casting after it leaves the boring machine.

FIG. 11. WHERE THE CYLINDERS ARE PLANED
In the foreground is seen the "electrical walking crane" which is used for transporting cylinders from place to place.

BALDWIN LOCOMOTIVE WORKS

The inner face of the saddle is planed up to match the face of the opposite casting, and the bearings for the front rails of the frame are also carefully machined. In the case of the cylinder now under notice, there are two front rails, one above and one below the casting. This work is done on large planers, in the shop shown in Figure 11. Jib cranes handle the castings, and there is also an electrical walking crane, shown in the foreground of the illustration, with which they are transferred from one part of the shop to another. A view of a large planer, finishing the bearing for a frame rail, is shown in Figure 12.

The casting is now prepared for receiving the cylinder heads, which are also of cast iron, and have previously been drilled with holes for bolting them to the cylinders. The heads are clamped in place, and the holes in the flanges of the cylinders are drilled through gauges to match exactly those previously formed in the heads. The seats for the heads are first carefully filed with a fine file, and then ground to a perfectly true face to insure a tight fit. The opening for the steam passage is finished with a concave seat to insure a tight joint, and the seat for the blast pipe is faced up and finished to a true surface. Holes are also drilled in the flange of the saddle, and the two halves are then assembled and bolted together.

It now remains to insert the steam-chest bushing before sending the cylinders to the erecting shop. The

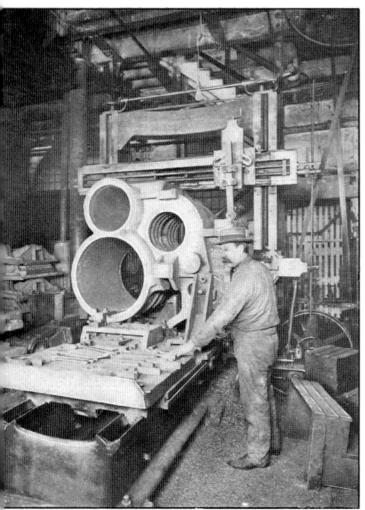

FIG. 12. A CYLINDER PLANER

This illustration shows the process of finishing the bearings for the frame rails.

BALDWIN LOCOMOTIVE WORKS

bushing is of hard cast iron, cylindrical in shape, with walls five-eighths inch thick. It has circular ribs to match those in the casting, and twelve longitudinal ribs, which stiffen it and give the valve a good bearing. The bushing is machined to size with great accuracy, in order that it shall properly fit the steam chest; the port openings being finished on a slotting machine, wherever they have not been formed the full size in the casting. The machine shop where these operations are completed is shown in Figure 13.

The bushing is now forced into the steam chest under hydraulic pressure by means of a special machine. The ram has attached to it a long rod which passes entirely through the steam chest and bushing, on the outer end of which is placed a cap, secured to the rod by means of a large nut. Pressure is applied to the ram by a small hand pump, and the bushing is slowly drawn into the steam chest, the maximum pressure, about forty tons, being reached during the last part of the operation. The bushing is drawn in until the circular ribs exactly match those in the casting. With this arrangement these important parts can be machined

FIG. 13.   MAKING THE STEAM-CHEST BUSHINGS

This illustration shows that part of the machine shop where the steam chest bushings are turned and the ports are trued by use of slotting machines.

with great accuracy; and when the valve seat becomes worn the bushing can be rebored, and replaced by a new one when it becomes too thin. The cylinders are now practically finished and are sent to the erecting shop for assembling with the other parts of the engine.

BALDWIN LOCOMOTIVE WORKS

The compound locomotive has, for more than ten years, proved itself an economical and reliable source of power in many branches of service. This is especially true of the Vauclain four-cylinder type, invented by a member of the Baldwin firm, and which is now in most extensive use on many of the leading roads at home and abroad. The Vauclain type dispenses with complicated reducing and intercepting valves, and presents an arrangement of cylinders which is exactly similar on both sides of the locomotive, and the power exerted on both sides is exactly the same. The mechanism is very simple, and as nearly like that of an ordinary single expansion engine as it is possible to make it. The compound locomotive has shown, in actual road work, a fuel economy over the simple engine amounting to from ten to twenty-five per cent. This is a matter of importance, especially where loads are exceptionally heavy and engines must be driven to the limit of their capacity to take them over the road on time. There are many large simple engines at work to-day in which the rate of fuel combustion must be forced to such a high figure that it is almost impossible to keep the required amount of coal in the fire-box; while similar engines have compound cylinders, with a milder exhaust and lower steam consumption per horse-power developed, burn less coal, are more satisfactory, and can, on a critical grade, keep a train moving where a simple engine would probably stall. No one realizes these facts more clearly than does the fireman, who, on many roads, has learned to fully appreciate the compound. One fireman, who runs on a road where the compound has largely supplanted the simple engine in fast express work, told the writer recently that "there was all the difference in the world between firing a compound and a simple engine."

The outlook for the compound locomotive is a bright one, and it is safe to say that, in the near future, it will more and more displace the simple engine in road work, where there are long hauls with heavy loads.

BALDWIN LOCOMOTIVE WORKS

Wheels, Frames and Other Parts

IT is customary at the Baldwin works, to use cast steel for driving-wheel centers on fast passenger engines. On freight engines either cast-steel or cast-iron wheel centers are employed, as may be required, and the new Baltimore and Ohio Consolidation locomotives have cast-iron centers, with the exception of the main drivers, the centers of which are of cast steel.

No steel castings are made at Baldwin's, being purchased from other firms; but all cast-iron wheel centers are made in the foundry, previously described in the paper on cylinders. A softer grade of iron than that employed in making cylinders is used. The space for the counterbalance weight is cast hollow, as is also the rim; which is split in four sections by narrow gaps, to avoid undue strains while cooling. The wheels, like the cylinders, are thoroughly cleaned before they leave the foundry; and are then sent to the wheel shop, a general view of which is shown in Figure 14.

The first operation is to prepare the wheels for forcing them on the axles. They are laid down on special machines with rotating tables, and the hubs are faced and then bored out to size with great accuracy. A keyway is slotted in each hub. The axles, of hammered steel, are in the meantime being turned and finished in lathes, all measurements being made to gauges to insure uniformity and accuracy. A keyway to match that in the wheel centers is cut at each end of the axle, these keyways being exactly at right angles to each other. The bore of the hub has a diameter less than that of the axle by an amount equal to three-thousandths of an inch per inch of axle diameter. For a nine-inch axle, for example, the hub will be bored to a diameter of eight and nine hundred and seventy-three thousandths inches. The axle is now hung between the uprights of a hydraulic press and one of the wheels set up in front of it, the two being so related that the keyways exactly match. The end of the axle and the interior of the hub are coated with black lead and oil, which acts somewhat as a lubricant when the wheel is forced on. The space between the

other end of the axle and the ram of the machine is filled up with metal blocks, and pressure is then applied to the ram and the axle is slowly forced into the hub under a maximum pressure (in a cast-iron wheel) of ten tons per inch of diameter. In forcing on a steel wheel a considerably larger pressure can be used without danger of injuring the center. After one wheel is on, the axle is turned around and the other forced on in the same way. Keys are then driven into the keyways at each end, and cut off flush with the face of the hub.

The wheels are now ready for turning, but before they are put in the lathe, the gaps in the rims, which have been previously machined, are filled by driving in wrought-iron filling pieces. The pair is now set up in one of the large lathes seen in the picture of the wheel shop, a more detailed view being given in Figure 15. These machines can be adjusted to fit wheels for different gauges, and are driven by separate electric motors. The face plates in the largest are 100 inches in diameter, and the wheels are rotated with the machine by means of heavy "dogs," which are clamped

FIG. 14. WHEEL SHOP

to the face plates. The centers are carefully turned up to the proper diameter to receive the tires, all the measurements being made to gauges.

The wheels are now taken out of the lathe, and the counterweights filled by pouring molten lead into the openings seen in the left hand wheel in Figure 15, until the exact amount required is obtained. Lead is also poured over the wedges previously driven into the gaps in the rim, thus giving a smooth finish at those points. The centers are now ready to receive the tires, which are made of a special grade of steel having great tensile strength. The tires, as they come to the works, are turned both inside and outside. An allowance of one-hundredth of an inch per foot of center diameter is made when boring out the inside of the tires. Thus, in the engine we are considering, the center is forty-eight inches in diameter, and the inside diameter of the tire is turned to forty-seven and ninety-six hundredths inches. The tires are heated in a furnace until they expand sufficiently to slip them easily over the center, a comparatively low heat being enough to increase their diameter by a full eighth of

FIG. 15. WHEEL AND TIRE-TURNING LATHE

an inch. A stream of water is then turned on them, and as they cool they shrink, and so bind themselves on the center without the necessity of using clamps of any kind. It is not customary, in domestic practice,

FIG. 16.   FINISHING CONNECTING RODS

is put on the center, the wheel runs sufficiently true after that operation is finished.

The final operation before the wheel leaves the wheel shop, is to insert the crank pins. These, like the axles, are of hammered steel, turned up in lathes, to fit the holes bored for them in the crank bosses. They are forced into their seats under hydraulic pressure, in the same way that the wheels are forced on the axles. It is unnecessary to use any special devices to prevent their working loose after they are forced into place.

The connecting rods of the new Baltimore and Ohio engines are of "I" section, fitted with the usual strap for holding the brasses, while the coupling rods are of rectangular section, with solid ends and plain bushings. The rods are of hammered steel, all the finishing being done at the works. The machine work is done almost entirely on planers and milling machines, the latter being most extensively used. In working on the body of the rod, a milling machine cuts a depression at each end, and the body is then finished to size on a large planer. These machines carry two tools and

to again turn the tires after shrinking them on, although foreign specifications are sometimes more particular in this respect. The tire having been turned up before it

BALDWIN LOCOMOTIVE WORKS

work on a pair of rods at the same time. In the case of an "I" section rod of uniform size throughout, a milling machine can be used to cut out the channels; but on a connecting rod which is made tapered, this cannot be done, and the milling machine is only used to cut a depression at each end, the rest of the channel being cut out by the planer. The bodies having been machined to size, the coupling rods are put on drill presses, and the holes for receiving the brass bushings carefully drilled out by means of special tools. The circular ends of the rods, and all large fillets, are finished on milling machines. In finishing an end, the rod is laid down on the rotating table of the machine, and so placed that the center of the crank pin hole coincides with the center of the table, and the tool thus cuts to an exact circle. The oil cups are forged solid with the rod, and are carefully machined and drilled with a special tool.

The bushings for the rods, after having been finished, are forced into place on a hydraulic press, which prevents their working loose and keeps them from turning. The straps and keys for the connecting

rods are carefully machined and finished by scraping, to assure an exact fit, and more or less hand-finishing work is done on the rods themselves, the end faces

FIG. 17. STEAM HAMMER SHOP

BALDWIN LOCOMOTIVE WORKS

FIG. 18. FORGING A FRAME

been used in the new Baltimore and Ohio engines. The castings are purchased; but the wrought-iron frames are forged up at the works from the raw materials. This process is a most interesting one, and we may well devote some space to describing it.

A general view of the hammer shop is shown in Figure 17, and a most excellent picture of a 6500-pound hammer, used for forging frames, appears in Figure 18. The frames are forged up from small pieces of selected wrought-iron scrap, which are first welded into thin slabs. A number of these slabs, piled together, are then welded and gradually worked into a frame; the top rail of which, in a large engine, is usually formed in two sections, which are welded together after the pedestals are forged on them.

The pieces are heated in a furnace conveniently placed with reference to the hammer. The fuel is in a separate chamber, and the hot gases, on their way to the chimney, pass through the flues of a boiler, thus raising steam with which to work the hammer. The piece being forged is slung from a crane, shown in the illustration, and handled by means of a large

being carefully scraped. A view of the shop where this work is done is shown in Figure 16.

Both wrought iron and cast steel frames have

BALDWIN LOCOMOTIVE WORKS

pair of tongs. The furnace is of sufficient size to contain two frames at the same time; and while one is being worked under the hammer, the other is being heated.

The top rail of the main frame is always forged, if possible, in one piece; but in the case of a large Consolidation engine, the frame is too long to be conveniently handled, and hence is forged in two pieces, which are afterwards welded together. It is the most recent practice at these works to forge the section for all four sets of pedestals in one piece, and then weld on the tail. In the case of the Baltimore and Ohio engines, however, the section for the second, third, and fourth pairs of pedestals was forged solid with the tail piece; and the section for the first pair of pedestals was then welded to it. This is called the "main weld," and it is not made until after the pedestals themselves are forged on and the middle and back braces welded in.

As the two sections for the top rail are forged into shape, the lugs for the pedestals are formed on them by means of suitable tools. A special gauge, on the end of a long rod, is used for making measurements of breadth and depth, the length being also worked to a gauge. The breadth and position of the lugs are determined by means of a sheet-iron templet. Another hammer, in the meantime, is forging the pedestals, which are provided with lugs projecting from them at right angles. These lugs have their ends notched like a V, and to them the middle braces of the frame are welded, as will be explained shortly.

The sections of the top rail and the pedestals having been finished, the latter are welded to the former. The lug on the rail and the top of the pedestal are first "scarfed" so that they fit properly. The under side of the lug is rounded to a convex surface, and a corresponding depression is formed on the upper end of the pedestal. The two are then brought to a welding heat, and pounded together under a small steam hammer.

The pedestals having all been forged on, the middle braces are cut to the proper length, and their ends pointed to fit into the notches cut in the lugs on the pedestals. The fit is an exact one, holding the braces, while cold, firmly in place. The joints are then brought to a welding heat, and the union completed by blows from sledge hammers.

The back brace is secured to the tail piece by an ordinary lap weld, its other end being "scarf welded" to the rear pedestal, like the middle braces.

The pedestals and braces being forged on, the next operation is to effect the main weld. The long section of the frame is notched, and the short section finished with a corresponding point, and the pieces are then laid together and heated in an open forge. When a high heat has been attained, the end of the frame is struck with a heavy iron ram, which effects a partial union and holds the sections together when they are removed from the fire. The main weld is completed on an anvil, by means of sledge hammers.

The frame is now a complete forging, in the rough; and during the remainder of the process cast steel and wrought-iron frames are treated practically alike. The sides are finished on enormous planers, which are large enough to work on a pair of frames at the same time, and take a continuous cut from one end of the top rail to the other. The pedestals are finished individually by taking short cuts across them, as time is saved in this way. All the inside work, such as finishing the lower surface of the top rail and the front and back faces of the pedestals, is done on slotting machines, the largest of which can work on three sets of frames at once. One of these machines, finishing two sets of frames, is shown in Figure 19. As will be noticed, there are practically two machines on one bed, each being driven by an electric motor, and independent of the other. This arrangement greatly facilitates rapid working.

The surfaces having been planed and slotted, the frames are sent to the shop shown in Figure 20, and here drilled on special machines, one of which is seen in Figure 21. This work is all done to gauges, thus rendering like frames absolutely interchangeable. The front rails are clamped to the main frames, and the two drilled together to insure an accurate fit.

The drilling having been finished, the frames are practically complete, and ready to go to the erecting shop.

BALDWIN LOCOMOTIVE WORKS

Many interesting processes are employed in manufacturing the various smaller parts which enter into the construction of the modern locomotive, but in this connection only a brief reference to some of them can be made. The various parts of the link motion are of hammered iron, case hardened for wear wherever there are sliding surfaces. The pistons are light castings, each being fitted with two cast-iron snap rings to prevent leakage. The rods are of iron, accurately finished and ground. The cross head is a steel casting, carefully machined. the wearing surfaces being covered with a layer of Babbitt. Tapered holes are provided for the piston rods, which fit into them accurately, and are held in place by a shoulder which is turned on them, a large nut being screwed on the outer end of the rod. The eccentrics are of cast iron, as are also the journal boxes. These are finished on large planers, a whole row of them being set up and worked together, thus assuring uniformity and rapid working. The springs are of cast steel, the material being subjected to special inspection, as its composition must conform to strict requirements. The leaves forming the springs are cut

to the proper lengths, and then assembled and held together in a clamp. The band surrounding the center is of wrought iron, which is welded to form a rectangle, and then heated. It is put over the spring while hot,

FIG. 19. FRAME-SLOTTING MACHINE

BALDWIN LOCOMOTIVE WORKS

FIG. 20. FRAME-DRILLING SHOP

and pressed into shape on a hydraulic machine. The shrinkage of the band, in cooling, holds the leaves firmly in place. The hangers and equalizers for the spring rigging are of wrought iron, forged into shape under drop hammers. An interesting detail is the forming of the steam dome covers used on these engines. The tops are hemispherical, and are formed of Swedish iron. This material is very ductile, and a flat sheet of it is worked over a suitable form by means of hammers, until it becomes a perfect hemisphere, with a smooth external surface. The dome casing is formed of three pieces, which are brazed together, and give a smooth outside finish. The various brass fittings are machined from castings made in the brass foundry. The metal is brought to the melting point in crucibles, which are placed in the furnace; for, unlike iron, the brass and the fuel cannot be mixed together in the same chamber.

A most important feature of the modern locomotive is the tender, the design and construction of which has recently been receiving a great deal of attention. The tender shop at the Baldwin Locomotive works is an extensive plant within itself, and some account of the construction of tenders will not be out of place here.

The Baltimore and Ohio Consolidation engines are equipped with tenders having 5,000 gallon tanks and wooden frames mounted on two four-wheeled diamond-frame trucks. The frames are built of oak and have heavy end sills, measuring ten by twelve inches, and four longitudinal timbers measuring six by twelve inches. Additional wooden braces run across the frame at its center, and at two intermediate points there are cross braces in the shape of heavy trusses, built up of bar iron, which carry the center bearings for the trucks. The timbers are held together by means of bolts and tie rods. The frame is covered with heavy planking, and the floor of the coal space is further covered with steel plates.

The tank is of the ordinary "U" shape, the walls of the coal space being vertical all around. Steel plate is used in its construction, the top and bottom each being made of three sheets, while six sheets are used to form the side walls. The rivet holes around

FIG. 21. FRAME DRILL

the edges of the sheets are punched on most ingenious machines, which automatically feed the plate, punch it, and at the same time shear the edge. The top and bottom of the tank are riveted to the side walls by means of angle bars, which are punched to the same pitch as the plates, on an automatic machine, thus assuring the fact that the holes will match. The angle bars are riveted to the top and bottom sheets by a hydraulic riveter. This is accomplished while the rivets are cold, the machine forming a flat head, and enabling the work to be done very rapidly. In the meantime the sheets for the side walls have, where necessary, been bent by means of machine-driven rolls, and they are now assembled with the top and bottom sections, and the whole united by a few bolts. As it is impossible to use the machine riveter after the tank is assembled, the remainder of the work is done by hand. The heads are round, and are formed on the outside of the tank in a suitable die. Half-way up the side suitable braces are provided all the way around, to keep the walls from bulging out owing to the pressure of the water. The riveting having been finished, the seams are calked with a round-nosed pneumatic tool, to render them water-tight. The tank is then filled, and again calked wherever leaks develop.

This work being completed, the tank is mounted on its frame, to which it is secured by angle irons; and it is then appropriately painted. There is one interesting feature in this connection. The capacity of the tank is plainly shown on the back, and on the front end of one of the water legs; while on the other leg is a scale showing the amount of coal in the tender. The scale runs up to 7,000 pounds, that being the capacity when the coal space is filled to the top of the tank, but not heaped.

The trucks, in the meantime, have been finished, and are now rolled into place. They have cast-iron M. C. B. journal boxes and double elliptical springs, and are braced with box girders built up of plates and angle bars, which carry the center castings; the truck bearings in the frame being set in heavy trusses, previously mentioned. After assembling the air-brake rigging and a few other details, and varnishing the tank, the tender is ready to leave the shop. It weighs about 100,000 pounds when loaded and ready for the road.

The next article will conclude the series, and will describe the process of erecting the locomotive after the various parts have been finished.

BALDWIN LOCOMOTIVE WORKS

Erection of Locomotive

THE erecting shop at the Baldwin Locomotive Works measures 160 feet wide and 337 feet long, and an excellent idea of its interior may be obtained from the accompanying illustration (Figure 22). The shop is divided, by the row of columns shown, into two bays, each of which is spanned by two Sellers electric traveling cranes of fifty and 100 tons capacity respectively. One of the 100-ton cranes is shown in the photograph, with a Mogul locomotive suspended from it. With the shop as at present arranged, there is space available for erecting about seventy engines at one time.

The accompanying illustrations are presented to show the progress during the erection of one of the Baltimore and Ohio Consolidation engines. The pictures do not all represent the same engine, but the locomotives are alike in all particulars, and the different stages are clearly shown.

The first thing is to set up the cylinders, which are placed on screw jacks at about the height above the ground that they will occupy in the finished engine. The upper surface of the saddle, to which the smokebox will be bolted, is chipped off by hand to insure a good fit. A line is scribed around the edge of the flange, and the casting chipped down to it over the surface surrounding the bolt holes; the seats for the steam and exhaust passages having been finished before the casting reached the erecting shop.

As a rule, the engine frames are set up before the boiler is put in place; but in the case now under consideration, the boiler was ready first, and was set up without waiting for the frames. This stage is shown in Figure 23. After the smokebox shell has been drilled to match the bolt holes in the upper flange of the saddle, the boiler is bolted to the latter, the cylinders being still supported by jacks. The firebox

BALDWIN LOCOMOTIVE WORKS

FIG. 22. THE ERECTING SHOP

As soon as the boiler is in place, the tubes are inserted. They are of iron, and have a copper ferrule at the firebox end, to insure a water-tight joint. The holes in the tube sheets are reamed out sufficiently large to enable the tubes to slip readily into place. By means of a special tool, operated by hand, the tubes are expanded on both sides of the sheet. The copper ferrule is omitted at the smokebox end, as the liability to leakage there is not very great; but at the firebox end, where the heat is more intense, it is of great service in preventing leaks.

The frames, in the meantime, have reached the shop, and are now swung into place. The two front rails, one of which is placed above the cylinders, and the other below, are bolted to the main frame, which is supported, at its rear end, by means of jacks. As the frame is the backbone of the locomotive, it is very important to have it accurately located. To assist in this a line is run from the center of the low-pressure cylinder exactly through its axis, and is secured to a standard placed beside the rear end of the frame. This is clearly shown in Figure 24, which represents the

is supported, on each side, by a jack set on a proper support. Some of the cab fittings are already in place, the cab being set, in this case, over the middle of the boiler.

appearance of the engine at this stage of the work. The frames are adjusted, by means of the jacks, until they are in exactly the right position when referred to the center line of the cylinders.

Attention is now directed to Figure 25, which shows a rear view of the engine after the frames have been leveled and the guides and guide yoke are in place. This illustration clearly shows the general shape of the firebox, which is of the "modified Wootten" type, without a combustion chamber. The outward flanging of the back head is noticed; also the location of the two furnace doors. The mud ring is single riveted, and the front is dipped toward the center. The frame for a steam gauge, placed so that the fireman can constantly note the pressure, is shown secured to the back head. There are many things about a firebox of this type which recommend it. A thin fire can be carried, without danger of blowing the fuel off the grate, very inferior coal can be burned successfully, and the evaporation per pound is increased owing to the more economical rate of combustion. A large firebox, with ample grate area, should mean

plenty of reserve power so far as ability to burn coal is concerned; and engines of this type can usually produce plenty of steam when being worked to the limit on the road.

Simultaneously with the inserting of the tubes,

FIG. 23. FIRST STAGE—BOILERS AND CYLINDERS ASSEMBLED

BALDWIN LOCOMOTIVE WORKS

more or less work is being done on the boiler in the way of drilling holes for bolting on the running-board brackets, air-pump frame, and other fittings. This work is done by hand, the tools being held by chains which surround the barrel. In the meantime, the

FIG. 24. SECOND STAGE—FRAMES IN PLACE

boiler has been secured to the frames by a heavy bearer in the form of a girder, which is bolted to the mud ring on both sides, at the middle length of the firebox; the front and back water legs being also secured to the frames by means of steel plates.

This stage having been reached, the next step is to get the driving wheels in place. The wheels, as they come to the erecting shop, have the journal boxes on the axles, the eccentrics being loose on the main axle. The wheels are placed on a vacant track near the engine to which they belong, and are spaced about the right distance apart. The boiler and frames are now swung over by means of one of the large cranes, and lowered down on the wheels; the clamps, which act as braces between the pedestals of the frame, being first removed. These clamps are hollow, and are held in place by a long bolt which passes through them and through the pedestal on either side, and is secured by means of a nut. The bolts are withdrawn and the clamps knocked out. The boiler is now lowered and the journal boxes guided in between their respective pedestals. The clamps are again put in place and the engine raised about a foot above the ground, and set upon jacks which are placed under the frame.

The appearance of the engine is now as shown in Figure 26. The guides and guide yoke are in place, and the crosshead is in the guides. The center bearing for the truck, a heavy casting, is shown bolted in between the front rails of the frame; the truck itself being one of the last things assembled before the engine leaves the shop.

By this time, nearly all the boiler fittings are in place. The throttle pipe, with its valve, is lowered through the opening in the dome ring, and secured to the dome by suitable braces. The dry pipe has fitted to its front end a casting with a conical seat, which is ground into a similar seat in the front tube plate, to insure a tight joint. The front steam pipes in the smokebox, which are of cast iron, are put in place, and the various washout and cleaning plugs inserted.

The boiler is now ready for the water test. The blow-off cock is connected to a suitable supply, and water forced in until the pressure is one-third greater

FIG. 25. SECOND STAGE—REAR VIEW

than the working steam pressure—in this case, to 267 pounds per square inch, the working pressure being 200 pounds. The leaks develop quickly, and the calkers work on the seams until the boiler is perfectly tight. The duration of the test depends upon the length

BALDWIN LOCOMOTIVE WORKS

FIG. 26. THIRD STAGE—DRIVING WHEELS IN PLACE

The seams and fittings are again carefully examined, and steam under full pressure is blown through the steam passages and cylinders. These tests are very carefully carried out, in order to prove, beyond doubt, that the boiler and all its fittings are in perfectly sound condition when the engine leaves the shop.

After the boiler has been blown out, preparations are made to run the engine before setting it down on the rails. The wheels are all "lined up," that is, adjusted to exactly the position they should occupy with reference to the center line of the cylinders. The connecting rods are set up, the link motion assembled, and the valves then set by means of trams until they show the amount of lead desired; after which the eccentrics are keyed to the main axle. Simultaneously with this work, the boiler is being lagged. Magnesia is used for the purpose, and it looks, from a distance, not unlike whitewashed boards. The boiler is encircled by wire hoops, to which the lagging is fastened by means of cleats. The back head is lagged down to the bottom level of the furnace doors—an admirable feature, and one that is becoming very com-

of time required to make the boiler tight. An hour usually suffices for this.

The water test having been finished, the boiler is emptied and tested with steam at 220 pounds per square inch, or ten per cent. in excess of the working pressure.

BALDWIN LOCOMOTIVE WORKS

mon in modern practice. The cab meanwhile is placed in position, it having been entirely finished and painted before reaching the erecting shop. The painting of the wheel centers, frames, and other parts, has been carried on simultaneously with the rest of the work.

The engine now appears as shown in Figure 27. It is noticed that the coupling rods are not up, they having been previously assembled and the wheels turned over to see that they fitted properly. The object of the running test is to see that the link motion and valves are in good condition, and that the cylinders and stuffing boxes are tight; and to ascertain this, it is unnecessary to have more than the main pair of wheels connected. By the time the engine is ready to be run, the sheet-iron jacket has been put over the boiler. The jacket, as received at the erecting shop, requires no further trimming, and each section fits accurately in its place. The cylinders are covered with cement lagging, and cased with a suitable iron jacket.

The boiler is now connected to a stationary steam plant, and filled with steam at about 180 pounds pressure. The engine is then given three running tests—

one with reverse lever in back gear, and two with the lever in forward gear. The performance of the engine is carefully noted, to make sure that the various parts are working properly. These facts having been ascer-

FIG. 27. FOURTH STAGE—THE BOILER LAGGED

FIG. 28.   FIFTH STAGE—THE ENGINE READY FOR SHIPMENT

jacks placed under the frames, the axles of the wheels can be dropped down between the pedestals, and the spring rigging set up without throwing any load on the springs. The oil cellars are set in, to replace the temporary wooden blocks which were used to line up the wheels before the engine was run. In the meantime, the driver brakes are being set up, and the air pump and injectors, with their piping and other attachments, put in place. Hand rails, cab fittings, running boards, and the many small details which help to complete the locomotive, are rapidly assembled. The grates, which are of cast iron, of the rocking pattern, are set on suitable bearers provided for them, and the ash-pan is then placed under the firebox. The truck, which is of the swing bolster type, with radius bar, is put under by raising the front end of the engine with one of the large cranes, and then rolling it in place. The truck wheels are raised by the crane, the center pin being guided into its seat, and are then propped up on wooden blocks, until the truck springs have been equalized with those over the front axle by a heavy equalizing lever, which has its fulcrum under

tained, the steam is blown off, and the shop tests thus finished.

The work of erection is now rapidly pushed to completion. As the weight of the engine is carried by

BALDWIN LOCOMOTIVE WORKS

THE ENGINE COMPLETE AND READY FOR SERVICE

BALDWIN LOCOMOTIVE WORKS

the cylinder saddle casting. The engine is then lowered once more, and the pilot bolted to the bumper beam.

The engine is now ready to leave the shop. Before putting on the lagging it was moved by the crane to one of the main tracks leading out of the erecting shop. After being hauled outside, the finishing touches are put on and a final inspection made, the engine appearing as shown in Figure 28. The locomotive, when shipped, is stripped of its connecting and coupling rods, smoke stack, headlight, and a few other fittings, which are packed in cases and assembled after it is placed on the tracks of the purchasing company.

The time required to erect an engine varies considerably, but, unless materials come in very promptly, it is usually at least five or six days after the boiler is set up before the engine is ready to leave the shop. The exact way in which the work is done depends somewhat, of course, upon the order in which material is received; but the description given above may be taken as fairly typical of the usual practice.

The finished locomotive, ready for the road, is shown on page 285. The two most conspicuous features of this design are the compound cylinders and wide firebox. The latter is unusually large for burning bituminous coal, the grate being nine feet six inches long by eight feet wide, giving an area of seventy-six square feet. There are 247 tubes, two and one-half inches in diameter and fourteen feet ten and one-half inches long, which provide 2148.3 square feet of heating surface; which, with 185.7 square feet in the firebox, gives a total heating surface of 2,334 square feet. The ratio of grate area to heating surface is thus as 1:30.7. This, when it is remembered that in some narrow firebox engines the ratio approaches and even exceeds 1:100, seems most unusual. The boiler is seventy inches in diameter, and the steam pressure 200 pounds.

The high and low-pressure cylinders are respectively fifteen and one-half and twenty-six inches in diameter, the common stroke being thirty inches. The driving wheels being fifty-four inches in diameter, the tractive effort, according to the Baldwin formula for compound locomotives, is 36,390 pounds. The total weight is 186,500 pounds, and the weight on drivers is

166,950 pounds. The tank capacity is 5,000 gallons, and the combined weight of engine and tender 286,000 pounds.

Taken altogether, the design is an interesting one, and the combination of a wide firebox and compound cylinders will doubtless give a good account of itself, especially when using an inferior grade of fuel. In the two particular features just mentioned, the design is quite in accord with the most up-to-date tendency in locomotive construction.

We have now traced, rather briefly, the progress made during the manufacture of some of the more important parts of the locomotive, as those parts are gradually converted from the raw material into the finished product. The length of time required to complete an engine varies largely according to circumstances and conditions of contract. Two to three months may, perhaps, be taken as an average. The shortest record at the Baldwin works is eight days, a narrow-gauge locomotive having been actually completed within that time. Perhaps the most striking feature connected with the work is the evidence that, in modern American locomotive construction, *utility* is the great object kept in view. Unnecessary finish is dispensed with, and the result is an engine that is strong, serviceable and practical, of moderate price, and able to do its work well until it is sent to the scrap heap to make way for something better. May the American locomotive continue, for all time, to maintain its high reputation, and to keep in the front rank of the motive power of the world.

In conclusion, the writer desires to express his thanks to Messrs. Burnham, Williams & Co., for kindly rendering every assistance in the preparation of this article.

Record of Recent Construction No. 30

BALDWIN LOCOMOTIVE WORKS

An Address upon

Compound Locomotives

To the Officers and Employees of

Union Pacific Railroad

Illustrated by Stereopticon Views   By S. M. Vauclain   Omaha, July 16, 1901

To the President and Officers of The Union Pacific Railway:

Gentlemen:—Complying with the request of your General Manager, Mr. Dickinson, I shall endeavor to give you some facts concerning compound locomotives, especially that type so well known to you all, and now in use on your railroad, the Vauclain four-cylinder compound.

Compounding, or the use of two or more cylinders for steam engines, is, as you are no doubt aware, always practiced when high duty and great economy are desired in both marine and stationary practice. It is thus resorted to, not because it is cheaper in the first cost, or because it requires less repairs, or is less complicated, and, I may even add, not because the class of labor required to operate it need be less skillful. Compounding was, and is now used, in both marine and stationary practice, for the reason that the many almost unsurmountable engineering problems of the past, have, by its introduction, been made easy. By its use the great ocean steamships are enabled to

BALDWIN LOCOMOTIVE WORKS

travel the high seas for long distances, and at the same time prove paying investments; the little torpedo boat is propelled with express speed, and with an efficiency which otherwise would be impossible; our huge battle-

ships can almost encircle the globe, and still have left resources sufficient to subdue an enemy. By it large manufacturing establishments have been converted from non-paying institutions into remunerative ones; boiler rooms and coal bins have been reduced so as to occupy the minimum amount of space, and huge and magnificent machinery is met with on every hand. No new enterprise can afford to undertake to do business requiring steam power by any other system. The question naturally arises: Why should not compounding apply to the locomotive? The time has arrived, the struggle has been a long one, but the slightly increased cost of the compound locomotive will no longer be an obstacle; its true value has been acknowledged, and its price is now the secondary and not the primary consideration.

Concerning the compounding of locomotives, the field was barren and bare scarce twelve years ago. It is true that much had been done by some inventors abroad to intro-

SLIDE 1. BALDWIN LOCOMOTIVE WORKS—From Broad and Spring Garden Streets

duce the principle on locomotives, but, as is frequently the case, the great desire for private gain on the part of the inventors had prevented extended trials of the many systems which were evolved. Many years ago in this country, men, who were much ahead of their day, developed and operated compound locomotives, but the defects of these were sufficient to engulf them, and their use merely created prejudice and made more difficult the way for the man of similar ideas to follow.

Your speaker kept in touch with all that was being done, both in England and on the Continent, but, in his judgment, the previous systems, such as those of Worsdell, Webb, Wolfe, Von Borries, La Page, Mallet, and others too numerous to mention, possessed none of the features necessary to make them popular, from an American point of view.

The requirements for success were as follows:

First. To devise a system of maximum simplicity, capable of showing a maximum economy, and at the

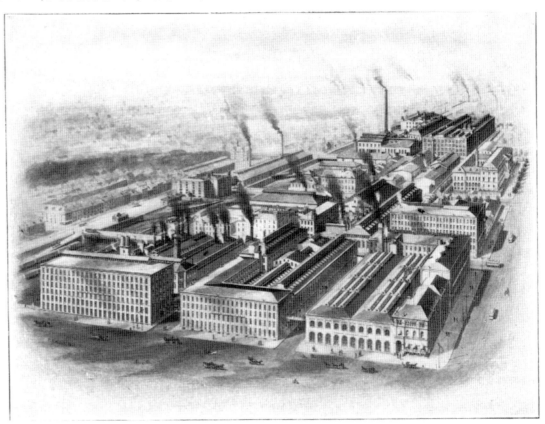

SLIDE 2. BIRD'S-EYE VIEW OF BALDWIN LOCOMOTIVE WORKS

same time applicable to all locomotives of any gauge, type or size.

Second. To be designed by an inventor who would be willing to sacrifice all hope of excessive gain in royalties.

Third. To be promoted by a well-established concern whose reputation for business integrity could not be questioned, and which would extend the world over.

The locomotive herewith illustrated by Slide 3, was the first Baldwin compound built, and even

SLIDE 3. FIRST BALDWIN COMPOUND LOCOMOTIVE

though handicapped by having only sixty-six-inch wheels, was used on the fast Blue Line service of the Baltimore and Ohio Railroad between Philadelphia and Washington. Your speaker was the inventor, and the Baldwin Locomotive Works introduced this type of loco-

motive and developed it until you have the present magnificent machines which haul your "Overland Limited." It must not be understood that we had no trials or tribulations. New cylinders were applied, and by the use of the indicator many of the imperfections of the first conception were overcome in those locomotives which rapidly followed. Many weary rides at night, on the front end of old "848," with an indicator and lantern, were necessary, but the data so gathered enabled us to proceed definitely and with continued success in the manufacture of compound locomotives.

The compounding of a locomotive, however, is not as easily done as that of a stationary or marine engine. The first and most important thing necessary is to provide a blast sufficient to generate steam, for without steam the engine would be useless, and, as you are aware, scarcely three feet of heating surface per horse-power is offered in locomotives, as against eight to twelve feet in stationary practice. Indeed, in our very fast eastern passenger service we are satisfied with one and three-quarter square

feet of heating surface per horse-power, if we have sufficient grate surface to enable us to burn enough coal to
do the work; therefore, a high rate of combustion is
necessary and is produced by contracting the exhaust
orifice, thus producing back pressure in the low-pressure

datum line is all dead work, opposing, as it does, the
advance of the piston. The higher the speed of the
compound the larger this loss becomes.

But this is true also of the single-expansion locomotive, and to a much greater extent, since it cannot

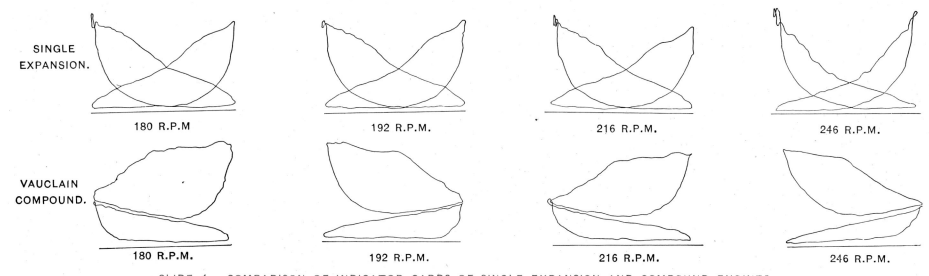

SLIDE 4. COMPARISON OF INDICATOR CARDS OF SINGLE EXPANSION AND COMPOUND ENGINES

cylinder. This can be readily understood by the diagrams shown by Slide 4, illustrated herewith, which are
exact records of the action of the steam in the cylinder.
The area included between the exhaust line and the

be run under similar conditions with as large an
exhaust orifice as can the compound.

Second. The cylinders must not extend beyond
the loading gauge of the road—they must clear the

BALDWIN LOCOMOTIVE WORKS

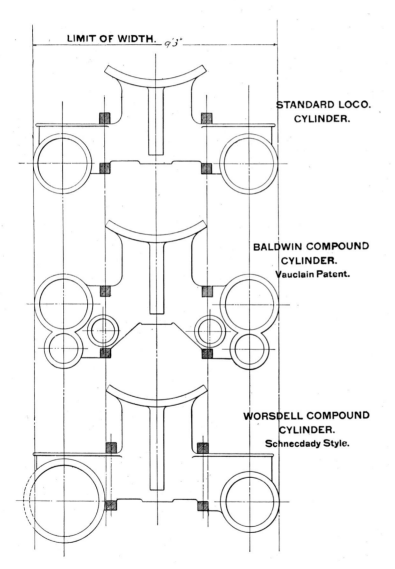

LIMIT OF WIDTH. 9'3"

STANDARD LOCO.
CYLINDER.

BALDWIN COMPOUND
CYLINDER.
Vauclain Patent.

WORSDELL COMPOUND
CYLINDER.
Schnecdady Style.

SLIDE 5. DIAGRAM SHOWING COMPARATIVE WIDTHS OF SINGLE
EXPANSION AND COMPOUND CYLINDERS

SLIDE 6. VAUCLAIN FOUR-CYLINDER COMPOUND, HALF SADDLE

track a reasonable distance—they must not interfere with the tracks. They must be so designed as to be interchangeable, rights or lefts as desired, and easily and cheaply applied. The diagrams shown herewith by Slide 5, will give an idea of the necessity of keeping within the width of loading gauge. It has now become impossible for the two-cylinder or cross compound to enter the field of large locomotives on this account, even though for this reason, the limit of width has been increased from ten feet to ten feet six inches.

The cylinder illustrated by Slide 6, on page 294, will give an idea as to how easily this can be accomplished. The saddle which joins the smokebox with its steam and exhaust openings, is exactly the same as

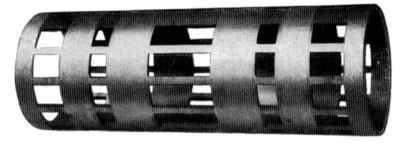

SLIDE 8. VALVE BUSHING

that used on the cylinders of single-expansion locomotives. The valve chamber or steam chest is merely a hole cast through the various walls which separate the many live and exhaust-steam passages. The high and low-pressure cylinders are cast one above the other and are connected to the steam chest, which is common to both, each of them being merely a round hole several inches long.

By reference to the diagrams illustrated herewith by Slide 7, there can be obtained a clearer idea of the manner in which the cylinders, steam chests and their passages are arranged.

Figure 1 shows the valve in the steam chest with the live steam, high pressure, low pressure and final exhaust ports laid bare, as well as the valve cavity or receiver.

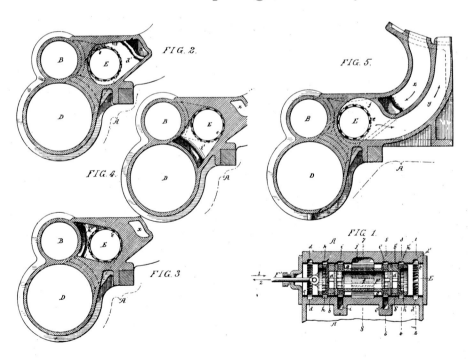

SLIDE 7. DIAGRAMS SHOWING CYLINDERS, STEAM CHESTS AND PASSAGES

BALDWIN LOCOMOTIVE WORKS

Figure 2 shows how the live steam passage is connected with the main pipe enclosed in the cylinder saddle and fed by the steam pipe from the boiler.

Figure 3 shows how the high-pressure steam enters the high pressure cylinder from the steam chest.

Figure 4 shows how the low-pressure steam is admitted to the low-pressure cylinder, and

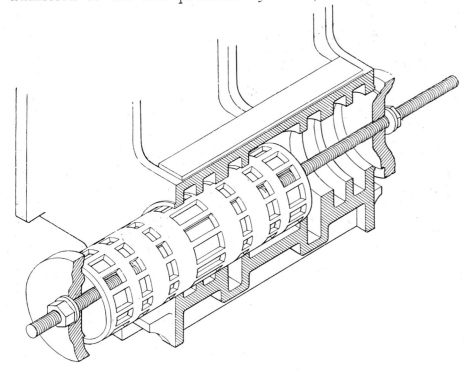

SLIDE 9. MANNER OF INSERTING NEW VALVE BUSHING

Figure 5 shows its final connection with the exhaust passage and the atmosphere.

The valve seat shown by Slide 8, on page 295, is, an independent bushing, machined accurately and easily to exact dimensions, and then forced into the hole in the cylinder by hydraulic pressure. By the use of this bushing repairs can be made from time to time; when desired changes can be made in port openings without altering the main cylinder castings. It also enables us properly to vent the cores of the mould in which the casting is made, and also to remove them from the casting after it has been cast. The figure illustrated herewith by Slide 9, will show how new bushings can be drawn into the cylinder casting when repairs are being made, where special tools are not obtainable. In fact this method was practiced by the Baldwin Locomotive Works for several years before hydraulic appliances could be perfected.

The crosshead, that part which has been most persistently attacked by those who would criticize the design of this machine, is merely a large lump of steel, properly designed, and strong enough to do the work demanded, and capable of withstanding any uneven strains which may come upon it. The guides are constructed likewise, and with experience as a teacher, it is felt that this feature is now thoroughly satisfactory.

Slide 10, illustrated herewith, shows the crosshead as ordinarily coupled with pistons, cylinder-heads, packing and guides attached.

Slide 11, on page 298, shows how the crosshead may be changed to suit Atlantic type locomotives or whenever it is necessary to shorten the piston rods.

Slide 12, on page 299, shows the crosshead coupled up complete, with exterior view of cylinders.

The success of this locomotive depends primarily upon the valve, which, with the utmost simplicity, controls the admission and exhaust to both cylinders as shown in Slide 13, on page 300. In addition to the ordinary functions of a valve, it furnishes sufficient receiver capacity to take care of the steam, which must be retained after the low-pressure admission has closed, until the high-pressure exhaust has closed. Thus, by

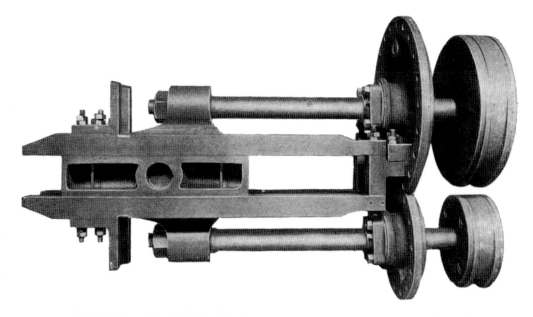

SLIDE 10. CYLINDER HEADS, PISTONS AND CROSS HEAD

proper proportioning of this detail, in connection with clearance in the high-pressure cylinder, we are enabled to keep the compression line of the high-pressure cylinder at or below the initial or boiler pressure, at the highest piston speed obtainable. This feature gave us more trouble than all the others combined, and was overcome only after much hard work and with the aid of our friend, the indicator.

I have with me a mechanical slide which, when

it is put in motion will enable you to understand better the functions of the valve. You will notice that the valve, at the same time, admits steam both to the high-pressure cylinder, through the live steam port, and to the low-pressure cylinder; this steam entering the low-pressure cylinder has in turn been exhausted from the opposite end of the high-pressure cylinder through the valve cavity. At the same time the valve allows the steam in the opposite end of the

low-pressure cylinder to escape through the final exhaust into the stack, having been sufficiently restricted at the final exhaust, so that it produces the necessary vacuum in the smokebox. This practice for steam-making purposes is common to any locomotive. This slide is illustrated, as far as possible, by diagram on page 313.

In order to start promptly it is necessary to allow live steam to wiredraw into the low-pressure cylinder, so that a somewhat equal pressure can be obtained on both pistons. This is accomplished by the use of a small three-way cock, marked "E" in the diagram shown on Slide 15, page 300, which is placed between the two ends of a small pipe. These ends connect the two openings into either the ends of the high-pressure cylinder or its admission ports. The lever operating it is attached to the cylinder cock lever, and can be used at the discretion of the engineer. It will be observed that when this cock is opened, steam will pass from the front end of the high-pres-

SLIDE 11.   ARRANGEMENT OF GUIDE AND CROSS HEAD FOR
ATLANTIC TYPE LOCOMOTIVES

sure cylinder through the pipe to the back end of the high-pressure cylinder, then through the valve to the front end of the low-pressure cylinder, and so do effective work. This device is used in emergencies with good effect, when surmounting heavy grades, but should never be used except in starting, unless the speed drops to five miles an hour and the locomotive is in danger of stalling. The ordinary cylinder cocks are used for the low-pressure cylinder, whereas the three-way cock drains the high-pressure cylinder. Pressure and air-relief valves are used, the former for water in low-pressure cylinders, and the latter to admit air to the cylinders when drifting. The air-relief valve for the low-pressure cylinder is applied ordinarily to a hollow valve stem, permitting the air to enter the steam chest of this cylinder direct—that for the high-pressure cylinder is placed in the steam passage leading to the high pressure ends of the steam chest.

The illustration shown by Slide 16, on page 301,

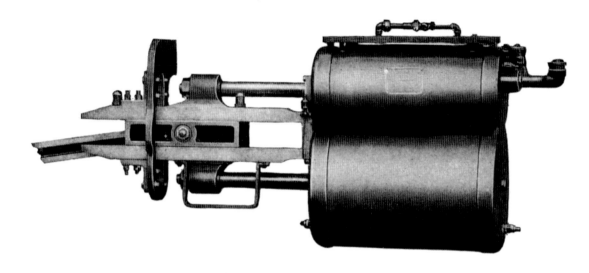

SLIDE 12.   CROSS HEAD COUPLED UP COMPLETE

will give an idea of the internal economy of one side of a Baldwin compound. The average locomotive engineer or fireman should not be too severely censured for not understanding the various détours which the steam must make, as from the outside nothing can be seen.

Now, gentlemen, as the construction of the machine has been made plain, I shall endeavor to give an explanation as to why a compound locomotive must save water and fuel over a single-expansion locomotive. The source of power is heat, and heat is produced, in this case, by burning coal in the firebox of the locomotive. The product of the combustion passes through the flues to the smokebox, and thence out of the stack, during which time a portion of the heat produced is imparted to the water in the boiler, thus generating steam.

The steam is used in turn in the cylinders, and the pressure or power is further transmitted to the crank pin, through the pistons and the main rod, producing a tractive effort dependent upon the pressure of steam per square inch in the cylinders and the dimensions of the cylinders and wheels. Is it not apparent then, that the engine using the smallest amount of water when doing

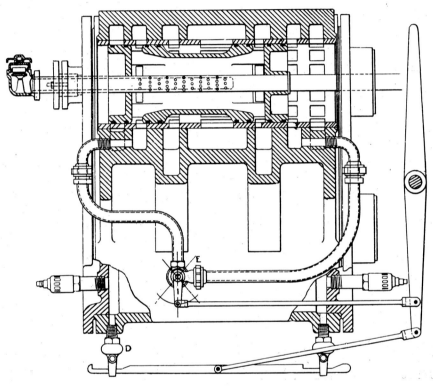

SLIDE 15. SECTION OF PISTON VALVE AND BUSHING

SLIDE 13. PISTON VALVE

equivalent work will use the smallest amount of coal? If the boilers are the same in both cases, the locomotive which burns the least amount of coal per square

foot of grate surface, will save an additional amount of coal per unit of work performed, on account of the slower combustion required.

The diagram shown by Slide 17, on page 304, will help to convey more clearly the extent of this economy. Owing to the high ratio of expansion in the compound, coupled with the greatly decreased condensation due to employing high-pressure and low-pressure cylinders, we are able to obtain in ordinary service a consumption of water at half stroke of only 14.7 pounds per horse-power indicated, whereas the single expansion locomotive, when at its most economical point of cut-off, uses 24 pounds.

The upper line will give an idea of how the water rate varies in a single-expansion locomotive, at various points of cut-off, for each horse-power of work performed, and the lower line the same for compound locomotives. You will also understand from it how, when on a dead heavy pull with the reverse lever down well in the corner, the compound locomotive goes along peacefully, confident of the result; while the single-expansion engine is thirsty for water all the time, and the poor fireman feels that the hilltop will never be reached. At three-quarter stroke we have the indicated water rate per horse-power of 16.2 pounds for the

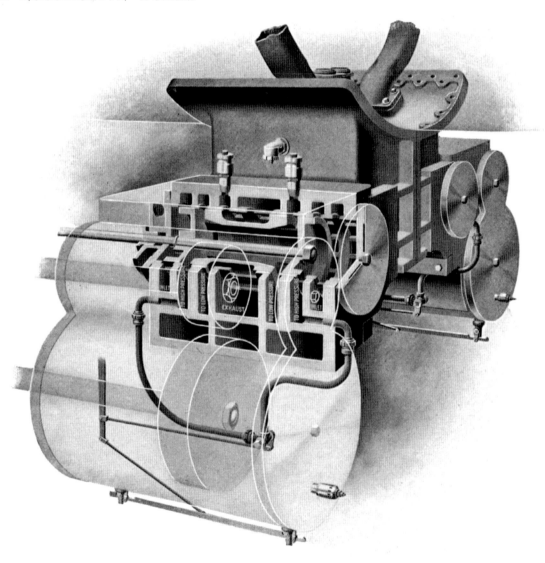

SLIDE 16. VIEW SHOWING GENERAL ARRANGEMENT OF CYLINDERS, WITH SECTION THROUGH STEAM CHEST VALVE AND PORTS

BALDWIN LOCOMOTIVE WORKS

compound, as against thirty-one pounds for the single-expansion engine, and if we add to this the water that is ordinarily consumed in other functions common to the locomotive, which in compound locomotives is not less than seventeen per cent. of the whole, and sometimes twenty per cent., we shall have an actual water rate at this point of about twenty pounds for the compound and thirty-five pounds for the single-expansion engine, or an economy of nearly forty-three per cent., which in part explains some of the enormous economies shown by the following tables:

ECONOMY OF FUEL
BALDWIN COMPOUND LOCOMOTIVE

Northern Pacific	26	per cent.
Western New York & Penna.	36.2	per cent.
Penna. & Northwestern	33	per cent.
Missouri, Kansas & Texas	36.6	per cent.
Norfolk & Western	38	per cent.
Western Railway of Cuba	27	per cent.
Philadelphia & Reading	33	per cent.
Cincinnati, N. O. & Texas Pacific	44	per cent.
Western Maryland	44.9	per cent.
Cornwall & Lebanon	33	per cent.

FUEL TEST
OF THE
NORFOLK & SOUTHERN RAILROAD
Conducted under Supervision of
MR. HERBERT ROBERTS, Superintendent of Motive Power

ENGINE No.	9	11	8	10
Cylinders (Diameter and Stroke) ins	12 20 x24	12 20 x24	18x24	18x24
Total Weight	105,000	105,000	94,000	114,800
Weight on Drivers	87,000	87,000	75,000	94,600
Heating Surface (total sq. ft.)	1274.8	1274.8	1287.5	1281.8
Grate Surface, sq. ft.	17.0	17.0	17.11	17.0
Steam Pressure	147.33	171.0	140.8	145.7
Number of Trips	4	4	4	4
Average number of Cars per trip	41.45	52.25	40.1	41.75
Running Time	20h 34m	25h 13m	19h 25m	20h 44m
Total Weight of Coal	33,500	31,50	40,500	46,300
Total Weight of Water	201,960	250,234	257,740	259,926
Boiler horse-power	289.6	287.7	385.0	363.0
Water per pound of Coal	6.0	7.94	6.36	5.61
Coal per square foot grate surface per hour	95.813	73.482	123.227	130.512
Square feet Heating Surface per horse-power	4.418	4.431	3.526	3.343
Coal per car-mile	3.35	2.92	5.32	5.06
Water per car-mile	20.16	23.1	33.84	28.41

Aver. Economy per cent. in Coal, per car-mile	39.57
Average Economy per cent in Water, "	29.69

BALDWIN LOCOMOTIVE WORKS

SUMMARY OF COAL TEST
M. K. & T. R. R.

ENGINE No.	235		245	
Weight on Drivers . . .	135,000		143,000	
Diameter of Boiler . . .	64 inches		66 inches	
Grate Area	25.19 sq. ft.		31.97 sq. ft.	
Total Heating Surface .	1768.3 sq. ft.		1739. sq. ft.	
Observer	A. Loucks	W. Maddocks	W. Maddocks	H. V. Wille
Train	Stock	Mdse.	Stock	Mdse.
Cars	31	28	29	32
Direction	North	South	North	South
Maximum Tonnage . . .	664.0	829.0	573.2	997.0
100 Ton Miles	1067.71	1046.93	886.726	849.86
Total Coal	11,130	12,200	15,300	18,200
Total Water	79,230	63,127	82,130	102,632.5
Coal per 100 ton miles . .	10.42	11.65	17.25	21.41
Water per 100 ton miles .	74.2	61.23	92.57	120.75
Water per pound Coal . .	7.12	5.17	5.30	5.64
Economy Water	19.8	49.3	. . . . . .	. . . . . .
Economy Coal	39.5	45.6	. . . . . .	. . . . . .
Terminals	Denison to Muskogee	Muskogee to Denison	Denison to Muskogee	Muskogee to Denison

Therefore, as coal must be burned to boil water, it is apparent that the compound locomotive will save coal, in exact proportion to the less amount of water used to perform each horse-power of work.

The term " horse-power of work performed " is used in order to make a definite comparison. Two engines hauling the same number of tons over the road can vary greatly in the amount of work performed viz.: horse-power hours developed. An unwise engineer may crowd the engine unnecessarily on the hills and work a full head of steam where the speed of his train would be ample under what is known as a cracked throttle. Another may take advantage of every opportunity to favor his engine, whilst making as good time over the road. Thus, two men can vary at least a ton of coal, hauling the same tonnage with the same locomotive. You can readily obtain the horse-power of a locomotive by multiplying the tractive effort by the speed, and then dividing by 375.

To the economy of coal, due to the less amount of water used, we must, however, add the coal saved by burning a much less amount per square foot of fire grate per hour—too rapid combustion is wasteful, and not only a much greater percentage of heat is wasted, but even unconsumed fuel is forced in large quantities through the stack. By overcoming this, additional economy is added to that already obtained by the use of compound cylinders exclusively.

BALDWIN LOCOMOTIVE WORKS

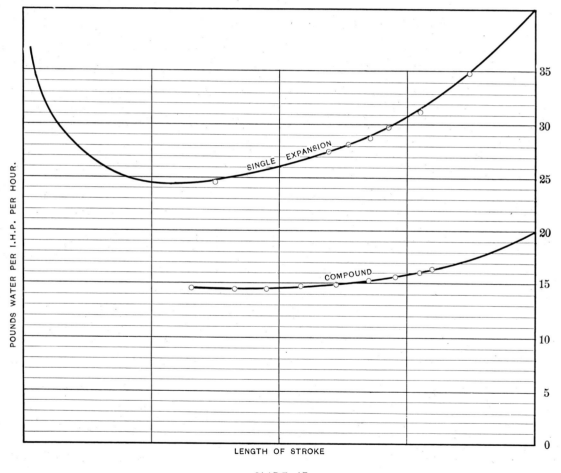

WATER RATE CURVES OF
COMPOUND AND SINGLE EXPANSION LOCOMOTIVES.

SINGLE EXPANSION

COMPOUND

POUNDS WATER PER I.H.P. PER HOUR.

LENGTH OF STROKE

SLIDE 17

The diagram shown by Slide 18, on page 305, graphically illustrates how the value of the coal varies as the rate of combustion increases. To secure this economy for single-expansion locomotives, and to still further increase the economy of the compounds, various roads are now extensively using wide fireboxes. For further information on this subject reference may be made to my addresses before the New York Railway Club and the Pennsylvania Engineers, copies of which can be had by writing to the Baldwin Locomotive Works at Philadelphia.

It is due to this last source of economy, that when heavy service tests are made, such high economies are obtained that sensible hard-headed men are prone to look upon the result with suspicion, and to attribute it to trickery or fraud. Such is not the case, however, and I feel that I have so clearly shown the reasons why,

BALDWIN LOCOMOTIVE WORKS

EFFECT OF VARYING RATES OF COMBUSTION.

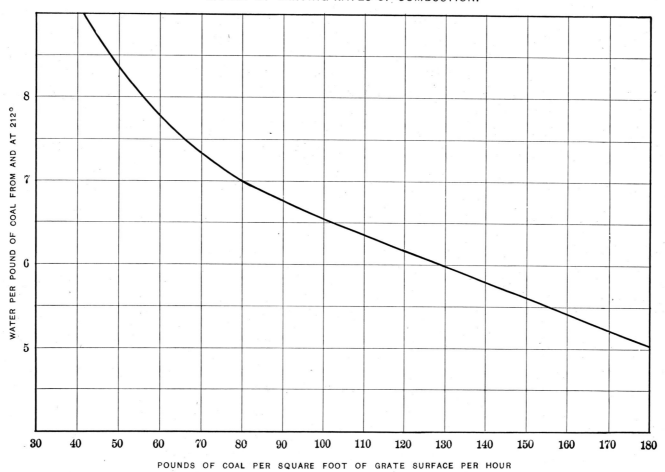

SLIDE 18

BALDWIN LOCOMOTIVE WORKS

that you will feel that some trickery is being practiced if you do not get such results from your locomotives. These compound locomotives are the only ones capable of producing constantly increasing horse-power as the revolutions increase, which is still another advantage. There seems to be in them no such thing as the "Critical Speed" mentioned by Professor Goss in his experiments at Purdue University, and fixed at one hundred and eighty revolutions per minute, for the Southern Express engine. Professor Smart of the same university, proved this lack of critical speed from a large working model presented to them by the Baldwin Locomotive Works, and the speaker also proved it by some high speed tests of locomotive 1027 on the Atlantic City Railroad. A view of this locomotive is shown by Slide 19, illustrated herewith.

The diagram illustrated by Slide 20, on page 307, will enable you to understand more clearly why the single-expansion locomotive, after reaching the critical

speed, fails to increase its horse-power, while the faithful compound goes on ever increasing until the limit of train speed is reached. It is for this reason that compounds of this type are so successful in high speed and in heavy passenger service—always maintaining their economy.

The diagrams illustrated by Slide 21, on page 308, were taken from a passenger locomotive at a speed of fifty-seven miles per hour. They are splendid examples

SLIDE 19. BALDWIN ATLANTIC TYPE LOCOMOTIVE

of hill work. No single-expansion engine could have been worked as was this locomotive, and that, too, without distress either to her fire or water. The maximum horse-power was 1429, and was obtained from a locomotive having only 1850 square feet of heating surface.

The excellent distribution of steam shown by Slide 22, on page 309, was obtained by using a longer cut-off than can be used in any other type of locomotive at that speed. The diagram will not only give the relative percentage of cut-off, but also the relative width of valve opening secured at the same speed, in both compound and single-expansion engines. The advantage the compound has over the single-expansion engine, in this respect, is accountable for many of the admirable results which are obtained in service.

In freight service, I need only exhibit some diagrams which were taken to show steam distribution

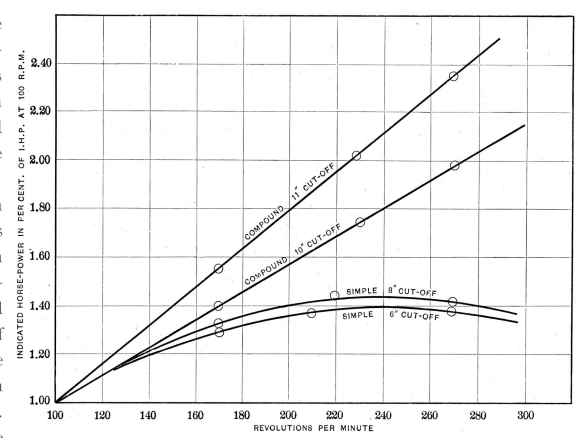

SLIDE 20. SPEED DIAGRAM

in one of our early locomotives. See Slide 23, page 310. These are not only good, but cannot be excelled. They account for the enormous economy in freight haulage, and are directly responsible for such locomo-

BALDWIN LOCOMOTIVE WORKS

tives as that shown by Slide 24, page 311, which can be operated satisfactorily only by using this system of compounding. Your present Superintendent of Motive Power, Mr. Higgins, was responsible for these engines, and the constant service they are giving the Lehigh Valley Railway Company fully justifies his advanced

In introducing compound locomotives on any railroad system, much depends upon the intelligence and progressiveness of the personnel. Therefore it must be pardonable for me to say that we never, in all our experience, have had greater satisfaction than has accompanied their introduction on your road. The

INDICATOR CARDS SHOWING UNIFORMITY OF POWER IN WORK GRADES.

A

B

C

D

I.H.P. 1429

I.H.P. 1382

I.H.P. 1370

I.H.P. 1287

SLIDE 21

engineering. To-day we are offering similar locomotives with five pairs of driving wheels, weighing 280,000 pounds, capable of a tractive effort of 62,500 pounds. These locomotives we guarantee one man will be able to fire easily and comfortably. They are illustrated by Slide 25, on page 312.

spirit of fairness has pervaded every department, and a surprising result has ensued. The percentage of broken parts, due to carelessness or ignorance on most roads, has, on this one, been zero. (Great applause.)

During the remainder of my remarks I will show

a cross section of my valve and cylinders with arrow heads showing paths of steam at one-third stroke, which may enable you to get a better understanding of the device. See Slide 26, page 313.

Now, to justify our judgment relative to the adoption of this system of compounding, as one applicable to all classes of locomotives, to various gauges and for all purposes, I will ask your indulgence.

We have now applied compound cylinders to the smallest as well as the largest locomotives in the world. We have applied them to locomotives of less than two feet gauge of track, as well as to those having a track gauge of 5 feet 6 inches. They have been applied to switching locomotives, double-end locomotives, American Eight-wheel type, Moguls and Ten-wheelers, also to Consolidation, Decapod, Atlantic and Prairie types; all doing well, giving high economy and great efficiency.

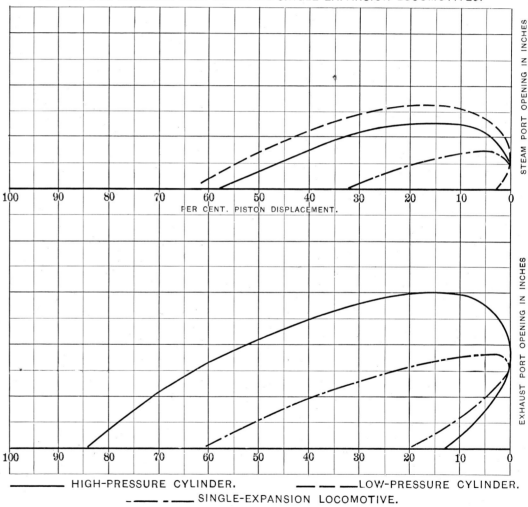

DIAGRAM SHOWING PORT OPENING FOR VARYING PISTON DISPLACEMENT AT RUNNING CUT-OFF FOR VAUCLAIN COMPOUND AND SINGLE-EXPANSION LOCOMOTIVES.

HIGH-PRESSURE CYLINDER.　　LOW-PRESSURE CYLINDER.
SINGLE-EXPANSION LOCOMOTIVE.

SLIDE 22

BALDWIN LOCOMOTIVE WORKS

Indicator Diagrams,
Baldwin Compound Locomotive.
Taken Dec 8th 1891.

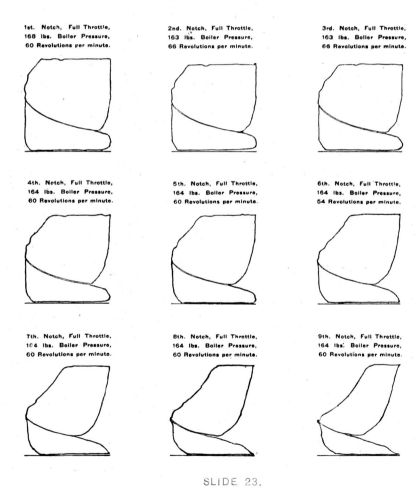

1st. Notch, Full Throttle,
168 lbs. Boiler Pressure,
60 Revolutions per minute.

2nd. Notch, Full Throttle,
163 lbs. Boiler Pressure,
66 Revolutions per minute.

3rd. Notch, Full Throttle,
163 lbs. Boiler Pressure,
66 Revolutions per minute.

4th. Notch, Full Throttle,
164 lbs. Boiler Pressure,
60 Revolutions per minute.

5th. Notch, Full Throttle,
164 lbs. Boiler Pressure,
60 Revolutions per minute.

6th. Notch, Full Throttle,
164 lbs. Boiler Pressure,
54 Revolutions per minute.

7th. Notch, Full Throttle,
164 lbs. Boiler Pressure,
60 Revolutions per minute.

8th. Notch, Full Throttle,
164 lbs. Boiler Pressure,
60 Revolutions per minute.

9th. Notch, Full Throttle,
164 lbs. Boiler Pressure,
60 Revolutions per minute.

SLIDE 23.

But this is not all—we have found this system of compound especially adapted to compressed-air locomotives, and already these busy little fellows are establishing a record.

On the steeper inclines, such as the Pike's Peak Railway, and many others in foreign countries, there can be found peculiar and wonderful locomotives compounded on this principle. The Pike's Peak Railway returned to us their three single-expansion locomotives, and had them changed into compounds at a large cost, because they could not afford to operate them at any price when compounds were obtainable. Fuel economy was the last consideration, the increased efficiency in making better time and an increase in safety being the reasons for the alteration.

At the present time compound locomotives are not only in use in almost every State in the Union, and on many of the leading and most progressive trunk lines, but their value has been recognized in every quarter of the world.

Even in a recent conversation with your president, he informed me that I had told the truth once, and that if results would show that I had told him the truth twice he would be delighted. I concluded that the first truth must mean that your compound locomotives had been economical in the use of fuel and as serviceable as the single-expansion engines. The other truth, which is still in doubt as far as I have been informed, must relate to my statement that the repairs to the compounds would not exceed those made to the single-expansion locomotives. I can confidently assure you, however, for your satisfaction, that this second truth will be proved. The Chicago, Milwaukee and St. Paul Railway has kept careful accounts to determine this very point, and they have reported to us that their compounds have shown about ten per cent. less for repairs than their simple engines. We do not promise any such result as this, although we have had it reported to us in several instances, but we do believe that, with proper supervision, the repairs to these locomotives will not be more than those necessary on single-expansion engines.

The cost of the maintenance of the compound cylinders will, no doubt, be more than that of the single-expansion cylinders, but the decreased demand upon the boiler, due to the less amount of water evaporated and fuel burned, per unit of work performed, will much

SLIDE 24. COMPOUND CONSOLIDATION LOCOMOTIVE, BUILT BY BALDWIN LOCOMOTIVE WORKS TO OPERATE ON MOUNTAIN GRADES

BALDWIN LOCOMOTIVE WORKS

decrease the cost of boiler maintenance, and probably more than offset, by the economy thus effected, the increased expense of cylinder maintenance. To illustrate this, let me state that some months ago I had been in service commencing with 1891, only two had had their tubes removed, and that recently and for the first time. These locomotives have indeed an enviable record, having over 500,000 miles to their

SLIDE 25. COMPOUND DECAPOD LOCOMOTIVE, BUILT BY BALDWIN LOCOMOTIVE WORKS

occasion to examine the flue records of the Chicago, Milwaukee and St. Paul Railway, and judge of my surprise to find that out of over one hundred compound locomotives in use on that road, which had credit, and are still in good condition. Twelve months is considered a fair length of time for tubes to remain in the boiler of single-expansion locomotives of competing lines operating under similar conditions.

BALDWIN LOCOMOTIVE WORKS

This may not all be due to compound cylinders, but may be due, in part, to the system followed by one company in taking care of their boilers, both of which, the cylinders and the system, must certainly be excellent and worthy of adoption by those seeking economical locomotive performance.

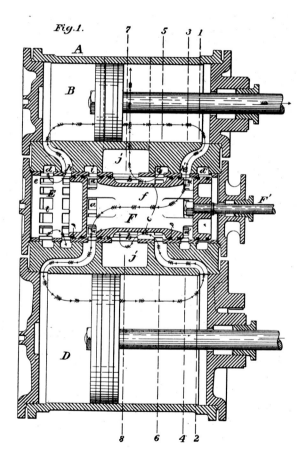

SLIDE 26. SECTION OF VAUCLAIN COMPOUND CYLINDERS AND VALVE

Index

BALDWIN LOCOMOTIVE WORKS

Index to Service

Index to Gauges

Index to Classes

Baldwin Locomotive Works

Philadelphia, Pa., U.S.A.

Builders of Single Expansion and Compound

Passenger Locomotives   Freight Locomotives   Switching Locomotives

Logging Locomotives   Plantation Locomotives

Locomotives for Rack Railroads   Locomotives for Mills or Furnaces

Heavy Locomotives for Special Service

Electrical Locomotives   Mine Locomotives   Oil-Burning Locomotives

Compressed-Air Locomotives

Specifications, proposals and full particulars furnished upon application